THE CENTRE FOR FORTEAN ZOOLOGY
expedition report 2010
India

Edited by: Tania Poole, Lisa Cole, Corinna Downes
Typeset by Jonathan Downes,
Cover and Layout by SPiderKaT for CFZ Communications
Using Microsoft Word 2000, Microsoft , Publisher 2000, Adobe Photoshop CS.

First published in Great Britain by CFZ Press

**CFZ Press
Myrtle Cottage
Woolsery
Bideford
North Devon
EX39 5QR**

© CFZ MMXI

All rights reserved. Without limiting the rights under copyright reserved above, no part of this publication may be reproduced, stored in or introduced into a retrieval system, or transmitted, in any form of by any means (electronic, mechanical, photocopying, recording or otherwise), without the prior written permission of both the copyright owners and the publishers of this book.

ISBN: 978-1-905723-75-1

Dedicated to the memory of Rudyard Kipling,
but for whom...

CONTENTS

5. Contents
7. Foreword by Dr Karl P.N. Shuker
11. Something Lost behind the Ranges by Richard Freeman
27. Picture Section
115. My close encounter with the Mandebring by Rudy Sangma
119. Expedition Report by Adam Davies
121. My search for the Mandeburung by Jonathan McGowan
131. A miscellany of Indian Forteana by Oll Lewis
141. Epilogue: Disappointing News

FOREWORD
DR KARL P.N. SHUKER

Many different types of cryptozoological man-beast have been reported and sought after down through the years. Some, such as the various categories of North American bigfoot (sasquatch) and Himalayan yeti (abominable snowman), are familiar to most readers, whereas certain others are much less so, even within the cryptozoological community. One example that until now had traditionally fallen into this latter group is the Indian mandeburung – which is why I was delighted when in 2010 the Centre for Fortean Zoology announced that later that same year it would be sending out an expedition to the fairly remote Garo Hills within the mountainous northeast Indian state of Meghalaya in search of this relatively obscure (and highly elusive) crypto-primate, which is said to inhabit its subtropical forests.

Large, hairy, and bipedal, the mandeburung is reputedly similar to the human-sized version of the yeti, and has been likened by some locals and eyewitnesses to a sizeable terrestrial ape, but is it real – or just, as mainstream zoologists have tended to assume, a deceiving combination of misidentified known species, traditional folklore, and downright hoaxes? The CFZ team hoped to find out, by collecting as much anecdotal information as possible, and perhaps even bringing back some hard physical evidence. As you will discover when reading this report, they were not disappointed by what they did uncover on site – and that was not all either.

What always fascinates me with CFZ expeditions is that although they originally go out in pursuit of one specific cryptid, while in the field they invariably uncover information concerning a number of other, usually hitherto-unpublicised examples too, and this present expedition was no exception. As will be seen in this report, the team collected much information (and even an alleged bone) in relation to a very big, rooster-crested snake known there as the sankuni; they observed and obtained samples for analysis from the mounted antlers of what seemed to them to be a truly gargantuan form of muntjac (barking deer), larger even than the world's biggest known species (the Vietnamese giant muntjac, itself only discovered by science as recently as 1994); and they not only encountered but also photographed the skin of a lesser (red) panda allegedly shot locally during the 1960s, even though this species is not supposed to exist anywhere in India.

Moreover, a particularly interesting and informative bonus contained within this report is an appendix penned by Oll Lewis documenting a wide range of additional cryptozoological cases and Forteana reported in the past from India.

Once again, therefore, a CFZ expedition has engendered a fascinating and highly informative addition to the crypto-literature, in the shape of this report – within whose pages, incidentally, I was startled but delighted to discover that one of the Goro Hills' most enthusiastic native investigators of the mandeburung had been inspired as a youngster to pursue such mysteries by none other than one of my own books – namely, *The Unexplained*, first published in 1996. It's news like this that suddenly makes everything worthwhile!

Dr Karl P.N. Shuker, August 2011.

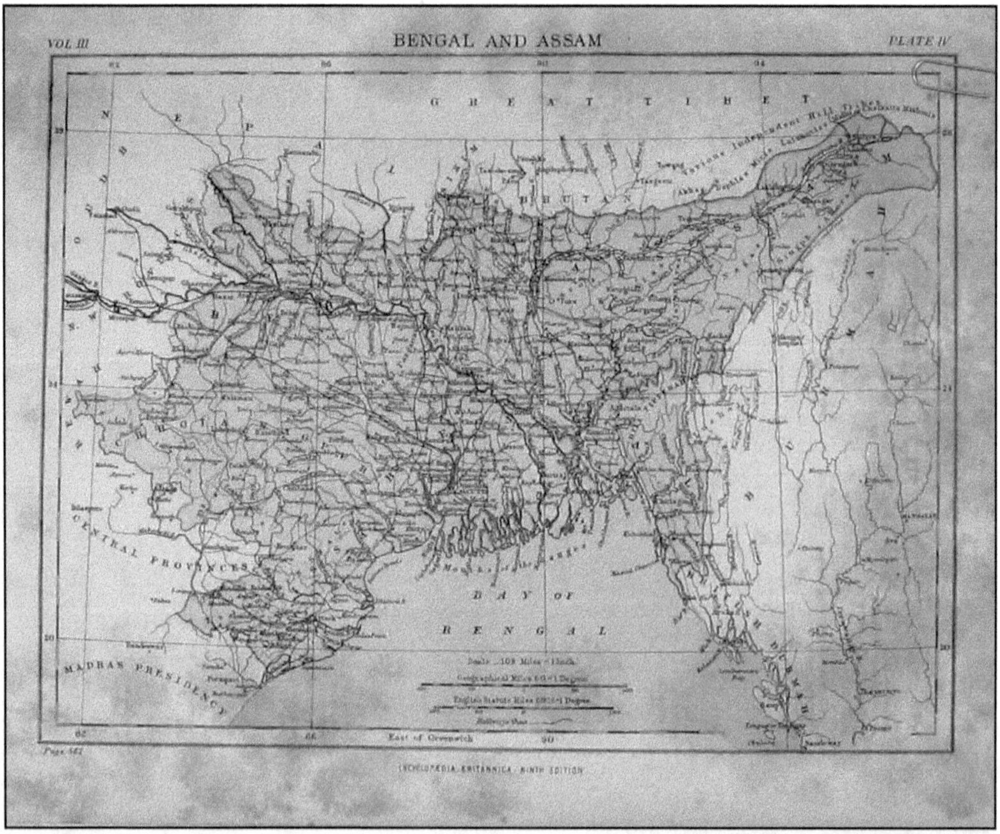

Meghalaya was formed by carving out two districts from the state of Assam: the United Khasi Hills and Jaintia Hills, and the Garo Hills on 21 January 1972. Prior to attaining full statehood, Meghalaya was given a semi-autonomous status in 1970.

SOMETHING LOST BEHIND THE RANGES
RICHARD FREEMAN

The expeditionary team of Dr Chris Clark, Adam Davies, Dave Archer and myself, who had previously searched for the Russian almasty (a relic hominid) and the puzzling Sumatran orang-pendek (mystery ape or hominid), were getting our heads together in planning where to go in 2010.

Several years before, Adam had been in Tibet on the track of the yeti. Ian Redmond, Tropical Field Biologist and Conservationist, mentioned to him that there were numerous reports of the yeti in the northern Indian state of Meghalaya. Upon returning to England, Adam investigated more closely and found that a local documentary film-maker and journalist, Dipu Marak had been on the trail of the creature for some years.

I, too, had heard of the Indian yeti, or as it is locally known *'mandeburung'* - the forest man. In June 2008 BBC journalist Alistair Lawson visited the area to investigate sightings of the creature. He was impressed by the remote, undisturbed landscape and wrote...

> "If ever there was terrain where a peace-loving yeti could live its life undisturbed by human interference, then this has surely got to be it.
>
> Perhaps the most famous reported sighting was in April 2002, when forestry officer James Marak was among a team of 14 officials carrying out a census of tigers in Balpakram when they saw what they thought was a yeti."

Dipu had given the BBC some hairs he had found at a remote area called Balpakram. Upon analysis, these proved to be from a species of Asian wild goat–antelope called the goral (*Nemorhaedus goral*). This, however, did not negate the eyewitness reports.

We decided that the CFZ team should investigate and began to lay plans for a trip to India. Adam, who is a great organiser, contacted Dipu who in turn planned guides, lodges and contact with eyewitnesses.

Jonathan McGowan was to join the four of us on this trip. Jon is an excellent field naturalist and taxidermist, as well as being the curator of the Bournemouth Natural History Museum.

EXPEDITION REPORT: India 2010

On Hallowe'en 2010, we flew out to India. During the long journey, we had to call for a doctor when Chris collapsed, but after he was given oxygen, he quickly recovered. The verdict was that he has been suffering from a lack of oxygen on the long stuffy flight.

We arrived in the mad cacophony that is Delhi in the evening and checked into our hotel. We had an evening to kill so we arranged for a taxi to show us some of the sights of the city. Unfortunately, the taxi driver just dumped us at a western style mall presumably thinking that, as westerners, it would be the place in which we were most interested!

Finally, we made it back to the hotel after avoiding wandering cows in the road and a near collision with a surprised Sikh in a three-wheeler. Driving in Asian cities is certainly an experience, which seems to consist of 90% horn beeping, and in India there is even a horn code, whereby certain numbers of beeps mean certain things. The legend 'Horn Please' is amusingly written on the back of many vehicles.

The following day we flew out from the surprisingly clean and efficient Delhi Airport to Guwahati in Assam. We were met at Guwahati by our chief guide Rudy Sangma, along with our assistant guide Pintu, and our drivers. We then began the long journey to the town of Tura in the West Garo Hills.

Meghalaya is a mountainous state in the northeast of India. It was carved out of the state of Assam in 1972 to accommodate the Khasi, Garo, and Jaintia tribes who at one time each had their own kingdoms. The three territories had come under British administration in the early 1800s and were assimilated into Assam in 1835. Once fierce head-hunters, the Garos were among the first Indians to be converted to Christianity by British missionaries. After conversion, the tribes were largely left alone, allowing a lot of their culture to remain intact.

This expedition was to be somewhat atypical. Generally, we camp out 24/7 in the jungle, mountains, desert or wherever, returning only to 'civilization' to stock up on supplies. This time, however, the Indian Government would not let us stay overnight in the jungle due to the activities of the insurgent group the Garo National Liberation Army. This significantly reduced our chances of seeing the mandeburung.

As the winding roads rose upwards, giving way to rocky tracks Rudy told me of some of the other strange creatures from the folklore of the Garo Hills. One creature that looms large in the Garos is the *sankuni*. This is a monstrous snake that bears a crest upon its head much like rooster's comb. Sounds familiar? It should do - the description of the sankuni matches up very well with that of the naga (the vast crested serpent I searched for in Thailand back in 2000) and the ninki-nanka (the serpent dragon of the Gambia I hunted in 2006). All are said to bear crests, be of huge size, have shining black scales, live in lakes or rivers as well as subterranean burrows, and to have an association with rain. This uncanny dovetailing of these stories made me seriously wonder if the sankuni and other monster snakes are based on encounters with a real-life species of gigantic snake unknown to science. Unlike the naga or ninki-nanka, the sankuni is also associated with landslides. Its underground crawling is supposed to cause massive shifts in wet earth. This sounds much like the weird South American serpentine cryptid known as the minhocão that is said to cause disruption, uproot trees, destroy houses and even alter the course of rivers. The sankuni is not wholly malevolent. Indeed, in legend, it is said to allow humans to use its great coils as a bridge allowing them to cross rivers. It is also said to manifest in dreams warning people of impending landslides. The sankuni is said to crow like a rooster much the same as the crested crown cobra of Africa. Its likeness to both the European basilisk (save in vast size) and the giant serpentine lindorms and worms hardly needs to be stated.

Rudy told me of another weird entity from folklore: the *skaul*. This is a vampiric entity that resembles a normal human being by day but at night its head detaches from its body and flies about as an independent entity. It has luminous hair and saliva. The skaul is said to feed on excrement and rubbish but also to suck up the human life force causing the victim to fall ill, weaken and finally die. The skaul may have been an early attempt to explain disease and illness. The luminous hair and saliva might well be based on early sightings of ball lightning or some other meteorological phenomena. The skaul has analogues across Asia with the Malayan *Penanggalan*, the Philippine *Manananggal*, the Balinese *Leyak*, the Thai *Krasue* and the Japanese *Nukekubi*.

We arrived in the ugly mountain town of Tura. Soulless grey buildings sprung up like crowded fungus, whilst gas and water pipes snaked above ground like rusting metal mycelia. The town was dirty, noisy and smelly. We checked into the tumbledown *Sandre Hotel* and unpacked.

Our meeting with Dipu Marak, the man who had been on the trail of the mandeburung for many years, offset Tura's unappealing nature. A delightful man, Dipu has a deep and infectious passion for the Indian yeti. He told us how he recalled hearing stories of the beast in his childhood and how they had sparked his lifelong interest. With a Garo mother and a Bengali father, Dipu is a huge fellow who towers over everyone else in the town.

The native Garos are quite distinct from the average Indian. They are an oriental race who originated in Tibet. They fought their way down to India and finally settled in the hills that bear their name to this day. Property and land are passed down the female side of the family, a wise move in a people who had to fight every step of the way on their long migration.

The following day we journeyed to Nokrek National Park. The hills here are covered by deep virgin rainforest. It was here that we intended to leave our camera traps for the duration of the expedition. The wild ancestor of all modern varieties of orange was discovered in these hills, and I sampled some *Citrus indica*, finding it to taste like a less sharp lemon. Another plant growing in abundance was a small, dirty brown, spherical fruit the locals called 'tastecan'. They looked like oversized oak gall but tasted exquisite. The flavour was quite unlike anything I have ever tried before, and to attempt to describe it would be akin to trying to describe a new colour. Rudy told us that, a few weeks before, the area was swarming with elephants and wild buffalo but they had now moved on. We heard hoolock gibbons calling in the distance.

EXPEDITION REPORT: India 2010

We took an arduous trek into the rainforest, where the terrain was very hilly and consisted of constantly climbing up and down ridges. We came across the nest of a wild boar and climbed down a dangerously steep cliff to investigate a small cave. The cave offered up no results other than the paw prints of a small felid, possibly an Indian leopard cat.

We planted camera traps at several locations making sure each had a good view of the area. All the traps were baited with bananas and oranges. The mandeburung is supposed to be primarily a herbivore although there are a couple of sightings of the creatures eating freshwater crabs. Dipu told us of one case were a farmer saw a family of four mandeburung stealing pineapples from his fields. The creatures ran away upon seeing him, snatching fruit as they went.

We moved from Tura, in the West Garos, down to Siju in the South Garos where we were met by Rufus - a friend of Rudy's - and another guide. We stopped in a rather down at heel and basic, but clean, tourist lodge. Close by were the Siju Caves where the village Headman had supposedly encountered a mandeburung several years before. The whole area was awash with wildlife from Indian false vampire bats and tokay geckos in the kitchen, to tarantulas on the walls outside. Jon McGowan used some fishing line and a live cricket to go tarantula fishing, baiting the spider out far enough to enable it to be photographed.

The caves themselves were amazing. Apparently, they go for miles with many smaller passages branching off the main cave. Fulvous fruit bats roosted in the cave and bizarre white fungus sprouted up from their droppings. The waters that ran through the cave were alive with tiny fish, shrimp, crab and cave crayfish. A swarm of them were feeding on a dead bat. Jon found two recently demised bats and decided to take them back with him to be stuffed. Huntsman spiders as broad as an adult human hand scurried over the rocks. I was excavating in the earth of the cave in the hope of finding some bone material as Pintu, one of the guides / porters, found a section of what looked like leg bone under some rocks. It was around six inches long. Upon examining it in the daylight, Jon thought it looked like the femur of a biped. We kept it for analysis.

The following day we set out across a huge suspension bridge that spanned the Simsang River and began to trek into the jungle. Early on Chris complained of pains in his chest and turned back, leaving us all quite worried. Despite being the eldest among us, on previous expeditions he had romped up mountains, across deserts and through jungles - treks that had left the rest of us gasping. It was clear that there was something wrong with him.

As we entered the jungle, a huge Bengal eagle owl went crashing through the canopy. As the path rose we glimpsed wild jungle fowl, the ancestor of the domestic chicken. This place really did remind me of Kipling's India. We came across an area of limestone outcroppings in the jungle. Some had been sculpted by wind and water to resemble human faces; others looked like the walls of lost temples or ruined cities, though all were natural in formation. They brought to mind the 'Cold Lairs' were the *'Bandar-Log'* or monkey people brought Mowgli in *'The Jungle Books'*. One in particular was a narrow passage between two limestone cliffs, and Rudy and Rufus told us that up until around 20 years ago the passage was used by hunting tigers to ambush men. Humans were forced to walk single file and the walls were too steep and slippery to climb making the men easy prey for the great cats. Later we came upon a watering hole and searched the mud for tracks; we found elephant, sambur, barking deer and buffalo. At one point, as we were resting in the jungle, something leapt down from the trees just over a ridge above us. The guides thought it may have been a leopard that was stalking us but on examination they said it was more likely to have been a monkey, and though we saw none we did find many monkey droppings.

The paradox of the jungle is that although it contains the greatest concentration of life anywhere on Earth animals are more difficult to see here than anywhere else. Creatures can hear a human coming from a long way off and melt like ghosts into the shadows. Wildlife is much easier to spot in open grassland areas. In all my time in many rainforests around the world I've only seen a handful of large animals.

Whilst most of us had been away in the jungle, Dave Archer had stayed by the Simsang searching for snakes and looking for animal tracks. He had found the footprints of a tigress in the sand - it was good to know that there were still tigers in the area.

Later we interviewed the Headman of the village, Gentar. He had encountered something strange in Siju caves several years before, something that had frightened him so much that he refused to go back there. He and some friends had been fishing by the light of burning torches. They had heard a noise that he described as sounding like someone treading on bamboo. On investigation, they found wet footprints on the rocks, which were human-like but of a vast size. They led down one of the passages that turned off the main one. The group thought a mandeburung had entered the cave from one of its many jungle entrances, and they panicked and fled the cave.

I found it odd that such a creature would be lurking so close to human habitation but I was to hear subsequent stories of them approaching other villages. Cave systems retain a stable temperature, it could be that the creature had entered the caves to keep cool or possibly to hunt for crabs.

Rudy and Rufus told us of their worries over the future of Garo culture. The younger generation are losing interest and increasingly wanting to become westernised. Only the very remote tribes are still animist and still hold onto all the old beliefs that are beginning to die out elsewhere, and the two are planning to write a book recording Garo culture and custom before it is lost.

The lodge we were staying at in Tura was supposed to be haunted. The first room on the right was the one where odd things were supposed to happen. Rudy told us that people sleeping in there had reported the bed shaking and shadowy figures. None of us were sleeping in that particular room, but when I looked at it if felt fine to me. The kitchen however had a positively malevolent feel to it. It was cold, nasty and held a feeling of 'something' watching you. Why the kitchen should feel so odd is anybody's guess.

EXPEDITION REPORT: India 2010

From Siju we moved down to Bagimara and set up HQ in a delightful lodge with a magnificent view of the Simsang. In the evening we would watch the sun setting over the river from the veranda. I enjoyed several chapters of Kipling's immortal *Jungle Books*, so cheapened and bastardised by Disney. Of all the places we stopped in India this was my favourite.

Whilst here we were introduced to a local man called Beka. A sculptor by trade, he had an interest in cryptozoology, and he told us of a story that his father had related to him. The event had taken place around 1940 in a lake near the borders of Bangladesh when a group of armed men, possibly soldiers, had shot a sankuni, which had, apparently, devoured a number of people over the years. The creature's body lay partly out of the lake and partly in, the portion out of the lake being said to measure 60 feet. I do wonder what kind of fire-arms would be required to cause such serious damage to a snake so huge, and also – if there is any truth to the story – what happened to the body? The story might be nothing more than a tall tale, but it does highlight the belief in a giant crested serpent in the Garos.

More recently - within the last five years – there was a case of a woman who dreamed that a man had warned her that her house would be destroyed, due to an impending landslide. Following the warning, she moved out of her house and it was destroyed as predicted. Witnesses saw a huge sankuni crawling away from the wreckage, so it could be that if the sankuni is a real, flesh and blood animal it inhabits underground burrows and lairs. If a landslide disturbs them, and the animal is seen crawling away, then people may well think that the sankuni's coils had been the cause of the landslide.

We travelled to the village of Imangri and trekked into the jungle beyond, where we saw simulacra of a footprint in limestone beside a river. It is a natural formation, but the fact it has been linked with the mandeburung argues that in order for such an association to have arisen, the creatures must have been known of for a long time. Swarms of yellow butterflies flittered around and we rested awhile beside the waters. Chris, had again felt ill that day and therefore stayed behind in the village.

We returned to Imangri and interviewed the Headman, Shireng R Marak - a 56-year-old with two thumbs on his right hand. In 1978, he and some friends were hunting in the forest, and as it was beginning to grow dark, he heard something big and powerful crashing through the forest. He heard a loud, deep call, which he imitated for us: AUHH!-AUHH-AUHH! He had heard village elders talking about the mandeburung and demonstrating the sound it made. He and his friends ran into a cave and lit a fire at the entrance, and they heard the creature bellowing and crashing around outside the cave all night. At first light it moved away into the forest and they ran back to the village.

Shireng said that 40 years ago sightings of the creature were more common. His friend's grandfather had shot one. He said it was man-like, covered in black fur and had a face like a monkey.

EXPEDITION REPORT: India 2010

A boy was sent to the fields to look for the village shaman who had also supposedly seen the mandeburung, and in the meantime I took a short trip by canoe down the Simsang River. When the shaman had returned to the village we interviewed him over tea.

Now suffering from cataracts, Neka Marak is 77 and once made medicines and charms. Back before the Indian/Pakistan war of 1965, he had been searching for an incense tree in the jungle when he came upon some thick creepers that had been snapped by something with immense strength. Hearing a crashing sound, he turned around to see a huge mandeburung charging at him through the jungle. To give us an idea of the size of the creature, Neka pointed to the roof of a nearby teahouse in the village. I don't know if it was the old man's cataracts making him over-estimate, or whether it was the length of time that had elapsed since the sighting, or even perhaps sheer fear, but the roof was 15 feet off the ground - a size I was totally unable to accept for the mandeburung. Neka went on to say that it resembled a huge hair-covered man, with hands that were big enough to have broken a human's neck. He said the face looked very human, but after all this time, he could not recall the colour of the creature's fur. To his credit, he did not try to embellish, but admitted that he could not recall the colour of the hair. Upon seeing the creature, he had fled from the forest as quickly as he could.

Neka had also seen the sankuni sometime prior to 1965. He saw the creature emerge from a cave beside the Simsang River, but he did not see the whole animal as he beat a hasty retreat. He indicated that the portion he saw was in the region of 25-30 feet long. It was black scaled with a yellowish underbelly and had a red, rooster-like crest and red wattles under the lower jaw. Neka had fled in terror from the giant snake.

The following day we drove to Balpakram, which is an area that looms high in Garo legend due to it being thought to be the place where the souls of the dead rested before going into the next world. It is a national park, and the forested areas are full of wildlife.

The roads grew more treacherous as we drove higher, and soon even the four-wheel drive vehicles were struggling to cope. We walked the final couple of miles on foot to the great plateau that formed Balpakram. I noticed that the area was heavily used for grazing and there were quite a few people around. Herdsmen were burning off the dry grass to promote new growth for grazing their livestock, and I found it hard to visualise a large ape, or indeed any big animal, existing in the area. The basalt rocks in the park were formed into six-side geometric forms much like those in the Giant's Causeway in Ireland. The molten rock formed the shapes as it cooled and contracted, although - unlike in Ireland - there are no columns and the shapes are visible only at ground level. Local people call the strange configuration the 'ghost market'. Rudy told us that fossil pumpkin, melon and tomato seeds have been found in the area, leading to the legend that it is a place where spirits hold a marketplace at night.

As we walked further across the plateau we finally came across a truly spectacular gorge. Seven kilometres wide, two kilometres across and around one kilometre deep, the Balpakram gorge makes an astounding spectacle. Heavily forested, it has near sheer sides and a river runs through it. Rudy explained that the only safe way in was via canoe from a nearby village. Only two or so hunters ventured into the gorge per year and it was mostly unexplored. It looked as if it could easily hide a small group of yeti in its deep, inaccessible forests. Unfortunately, we did not have enough time to investigate the gorge, as such an undertaking would have taken a whole week. We made plans to return to the gorge on a future expedition.

Back at the lodge, we met the owner Bullbully Marak who told us how keen she was to promote eco-tourism in the area - the Garo Hills and Meghalaya, in general, are not often visited by tourists. Often mistaken for Indonesians or Malayans, Rudy and Rufus mentioned that they often feel like foreigners in their own country. The feeling throughout the Garos is one that the Central Government of India is ignoring them, and such feelings have lead to the formation of several insurgent groups in the area.

We returned to the somewhat depressing surroundings of Tura and met up with Dipu once more.

Cameraman, BAFTA winning director, and company founder Morgan Matthews and soundwoman Tara Nolan had made an epic journey to join us. Despite not having slept in nearly two days, they were keen to begin filming.

Tura itself is devoid of anything approaching nightlife. The one bar in the town was at the *Sandre Hotel* and closed at 10 pm sharp. The bartender seemed totally disinterested in making money and resented anyone who entered the bar after 9.45 pm.

The two hotels in Tura both had restaurants that were spectacularly badly run. Their menus were surprisingly varied, but most items on them were not available. This made ordering food a bit like the cheese shop sketch in Monty Python. Far worse than this, though, was the service. On one occasion, we ordered some soft drinks, and an hour later they still had not arrived despite three waiters standing around next to the fridge in which the drinks were. In the end, Dipu himself had to go and open the fridge and point the drinks out to them. On another occasion, I ordered soup and bread - the soup took an hour to come and the bread turned up an hour after that.

We were to spend the next day interviewing a number of people around Tura. The first on our list was Dr Milton Sasama, the Pro-Vice-chancellor of the Garo Hills University, who has written a number of books on the history and folklore of the Garo Hills. However, he does not believe in the mandeburung as he has never come across descriptions of the beast in any of his studies. He has only heard of the monster, like a giant orang-utan, in the past 20 years. He also asserted that there was no tradition of a yeti-like creature in Assam, the Indian state that lies between the Garo Hills and the Himalayas.

Conversely, he believed implicitly in the sankuni. He knew a man who had eaten the flesh of a dead, juvenile sankuni after it had been washed into a village by a flood. It was between 12 and 20 feet long and bore a rooster-like crest. The meat from the carcass had provided enough food for the whole village. The man concerned, now in his 80s and called Albin Stone, resides in Tura these days.

EXPEDITION REPORT: India 2010

Our next interviewee was Llewellyn Marak - the uncle of Rufus - who is a noted naturalist and author of a number of books on the wildlife of the Garo Hills. In 1999, he came across a set of four huge, man-like footprints at Nokrek Peak around 21 km from Tura. They were found beside a stream in sand and were 18 inches long, and led away into the jungle.

Llewellyn's grandfather was a renowned hunter who amassed a large collection of trophies. He had encountered the mandeburung on a hunting trip many years before, when he came across the beast in a jungle clearing. It resembled a huge gorilla and was black in colour, and moved around on all fours giving the impression that it was searching for food. Occasionally it would stop and sit, appearing to eat something. Llewellyn's grandfather became afraid and backed away. This is the only report we have of the creature moving on all fours, but then again it may have been doing this in order to forage for food. The experienced hunter was sure what he had seen was not a bear.

Llewellyn, a conservationist rather than a hunter, invited us to look at his father's collection. Eagle-eyed Jon McGowan spotted something unusual among them - a pair of muntjac horns of unbelievable size.

On closer examination, these very distinctive horns proved to be even larger than those of the giant muntjac *(Muntiacus vuquangensis)* of Vietnam and Laos. This picture shows the horns

next to those of the Indian muntjac *(Muntiacus muntjak)*, and the startling size difference is apparent. Local people have a name for this particular deer, calling it 'matchok'. We took some samples from the antler for analysis back in Europe.

Llewellyn had also heard stories of giant catfish and giant freshwater stingrays, much like those said to lurk in the Mekong River of Indo-China.

Following this interview, we moved on to speak to Rufus' uncle - a surgeon called Dr Lao. Dr Lao also believed that the mandeburung existed but he thought that it was now very rare. Dr Lao had a collection of books on Indian wildlife, and among them was a book entitled, '*A Naturalist in Karbi Anglong*' by Awaruddin Choudry, first published in 1993. The book, by one of India's best-known naturalists, records his time in the Karbi Anglong district of Assam, the Indian state to the north of Meghalaya.

One chapter of Chourdy's book is dedicated to the *khenglong-po*, a yeti-like creature seen in the area. As Assam borders onto Bhutan there is a link, or corridor if you will, directly from the Himalayas down to the Garo Hills along which yetis are reported, which totally refutes Dr Milton Sasama's assertion that no such creatures are reported from Assam.

He writes…

> "Singhason peak and some nearby areas are sacred to the Karbis. Here in the dense forest lives the Khenglong-po, the legendary 'hairy wildman'. The Khenglong-po is an important figure in the Karbi folk tale. Whenever I used to get reports of its existence, I dismissed them as fable or mistaken identification of an ordinary animal. But when the much experienced Sarsing Rongphar gave me a fresh report, I had to rethink. Sarsing had been my guide in parts of the Dhansiri Reserved Forest, and I found him to be an accurate and reliable observer."

Sarsing was a hunter who used dogs to sniff out game such as muntjac and porcupine, which he then dispatched with a long hunting knife. Even before his arrival a Karbi, Along Awaruddin Choudry, had heard of sightings of a large, bipedal ape. At first he asked witnesses if they might be mistaking the creature with a stump-tailed macaque (*Macaca arctoides*) or a hoolock gibbon (*Hoolock hoolock*) but the witnesses rejected this as they were familiar with both species. However, when his trusted guide told him of an encounter with the beast, Choudry was forced to change his mind.

It was on 13th May 1992 that Sarsing Rongphar and his friend Buraso Terang took Sarsing's hunting dogs into the Dhansiri Reserve Forest. In the afternoon, they came upon large manlike footprints that were around 18 inches long and 6-7 inches wide. The pair followed the tracks for 3 kilometres until their usually brave dogs began to panic. Fearing an elephant or tiger was close by they crept cautiously forward, and soon a loud breathing sound became audible - a '*khhr-khhhr*' sound. From 80-90 metres away they saw an ape-like creature leaning against a tree, apparently asleep. The witnesses were at a higher elevation than the creature and had a clear view due to the fact there was no dense undergrowth obscuring their vision.

EXPEDITION REPORT: India 2010

The creature was jet black like a male hoolock gibbon. It had thick bear-like hair on the body. The hair on the head was long and curly. The creature was a female with visible breasts. Its mouth was open and large, human-like teeth were apparent. The face, hands and feet were black and ape-like. In front of the creature was a broken tree and the hunters thought the creature had been feeding on it. They observed the sleeping animal for around one hour. Sarsing likened it to a giant hoolock gibbon but with much shorter fore-arms.

On reaching their village, they told tribal elders of what they had seen and were informed that it was a Khenglong-po, a kind of hairy Wildman that was thought to be dangerous.

Choudry took Sarsing to his camp and showed him pictures of the Asian black bear *(Ursus thibetanus)* standing on its hind legs, and the mountain gorilla (*Gorilla beringei beringei*). The hunter identified the latter creature as being a Khenglong-po whilst recognising the former for exactly what it was. Choudry interviewed Buraso Terang separately and got the same answers.

A Khenglong-po was once supposed to have wandered up the railway track from Langcholiet to Nailalung.

On another occasion Choudry talked to some hunters from Karbi Anglong in central Assam. They spoke of a large, herbivorous, ground-dwelling ape that they called *Gammi*. According to them two Gammis were seen together in 1982 feeding on reeds on the eastern slope of the Karbi Plateau in the upper Deopani area. An elderly hunter had encountered one in the Intanki Reserve Forest in Nagaland in 1977-78. The creatures are said to be covered in grey hair and to be man-like in appearance. The name Gammi means 'wild-man'.

Choudry concludes...

> "It seems possible to me that a terrestrial ape, larger than the gibbons existed in some remote parts of Karbi Anglong and adjacent areas of Nagaland. The creature was always rare and preferred the remotest corner of the jungle, and, hence, evaded discovery by the scientific world. Now with the forests vanishing everywhere, this ape perhaps faces extinction. Expeditions to the heart of the Dhansiri Reserve Forest and Singhason area may well produce some result. But for now, I am looking for any fossil evidence including skull, bone or part thereof. This will at least put the Khenglong-po at its right place, even if it is extinct. Lastly, if a large mammal like the Javan or smaller one horned rhinoceros (Rhinoceros sondaicus) can be discovered in recent years in a small pocket of the war-ravaged Vietnam, outside its known locality in Indonesia and beyond anybody's expectation, one cannot rule out a Khenglong-po in the forests of Karbi Anglong."

We can see, then, an unbroken link of yeti sightings from Bhutan down into India.

The following day we interviewed another witness. He was a 51-year-old teacher called Kingston. In 1987, he and a friend were on Tura Peak, when he saw large, five-toed, man-like tracks in wet sand beside a stream. The toes and heels extended far beyond his own and sunk an inch into the sand whereas Kingston's own tracks only sunk in half an inch. My size nines were bigger than Kingston's, but he told me that the creature's tracks were bigger than my feet. He also heard the mandeburung's cry, AUHH!-AUHH-AUHH! He imitated the sound, which was in line with that made by other witnesses. He had wanted to investigate further, but his friend was too afraid. Kingston added that he has heard the cry on Tura Peak since, within the last few years.

We took tea with Kingston and he told us of being bitten by a viper, from his description it was a white-lipped pit viper *(Trimeresurus albolabris)* and how he had used a 'snake-stone' to draw out the venom and save his life. The use of snake-stones is widespread in Asia, Africa and South America. They are not really stones at all but parts of animal bone that have been cut and shaped with sandpaper before being wrapped in foil and placed in a charcoal fire for 15-20 minutes. The porous bone is said to draw out the snake's venom. Kingston said that the snake-stone had adhered to the bite and 'drew up' all the venom from his arm before falling off. Studies have shown that snake-stones are nothing but a placebo peddled by quacks. Maybe Kingston was just very lucky and recovered from the injury naturally, or perhaps the viper did not inject the full payload of its venom. He told us the chemist in Tura still sold them, and we thought it might be nice to buy some as souvenirs - but as it turned out - snake-stones were not to be found among the modern medicines on sale at the chemist.

Later that day we visited the village of Apertee some 35 miles from Tura to meet a witness called Nicholas Sama. In the 1960s, he had seen the severed hand and forearm of a mandeburung at a village market. The forearm, which was being displayed on a store selling bushmeat, was as long as his whole arm. The hand looked like a man's but far larger, and the nails were long. The arm was covered in long black hair. Nicholas thought it was very old as the skin was desiccated. No-one knew from where it had originally come, but Nicholas knew what he was looking at was not the arm of a bear or a gibbon.

The next day we met with a most impressive witness in the village of Ronbakgre. Teng Sangma had heard that in April of 2004 a village carpenter had seen a female mandeburung suckling an infant in a bamboo forest close to Rongarre. He did not believe the story but then on the 24th of that month, he and a friend were hunting for jungle fowl in the forest when they came across a huge figure sitting with its back to them. Even in its sitting position it was five feet tall. It was covered with dark hair and had longer hair on its head that fell down onto the shoulders and the back. The shoulders were very broad. It was a female, and was suckling a youngster whose legs were visible at the side of its mother suggesting that the infant was sitting on her lap. The youngster was making gurgling noises. The adult was pulling down large bamboo stems and plucking off the leaves to eat them. The men got to within 50 feet of the creatures, watching them for 2 minutes before becoming afraid, when they backed away leaving the creatures, which – apparently - had not noticed them.

We explored the area, walking along a stream into the jungle. We found the tracks of a fishing cat *(Prionailurus viverrinus)* as well as what appeared to be barefoot human tracks. The latter were far too small to belong to an adult mandeburung but we tried to take casts of them just in case. Unfortunately, the ground on which they were imprinted was far too damp to make casting with Plaster of Paris possible. Prior to our leaving the UK I had looked into other mediums for making casts, but I could find no resin and was told by a DIY shop that Polyfilla was unsuitable. The only liquid rubber I could source came with a big, gun-like applicator that would be difficult to get through customs. We all agreed that a human probably made the tracks, but we filmed and photographed them anyhow.

The following day we attended the Wangala Festival on the outskirts of Tura. The festival, also called the One Hundred Drums Festival has its genesis in the pre-Christian tribal celebrations of the area. It was held in each village after the harvest. It is a "Thanksgiving" ceremony to Misi Saljong, also known as Pattigipa Ra'rongipa (The Great Giver) for having blessed the people with a rich harvest.

A ritual called the Rugala is performed by the Nokma (a village Chief) a day ahead of the Wangala,

> "...and in this ritual the offerings of the 'first hand special rice-beer' along with cooked rice and the vegetables are given to Misi Saljong, the Giver. On the next day, the Nokma performs Cha'chat So'a ceremony or the burning of incense at the central pillar of his house to mark the beginning of the weeklong Wangala Festival."[*]

With the influence of Christianity the festival began to die out in all but the most remote villages. So in order to protect and preserve, and promote this culture's identity, a group of Garos decided to organise the Wangala Festival along modern lines. A group of 30 dancers with ten drums would form a contingent, and ten of these contingents - 300 dancers - would make up the Hundred Drums Wangala Festival. It has been held every year from 1976.

We were lucky enough to be guests at this gathering and to meet a number of local dignitaries and try some of the locally brewed rice beer. The performers, who had come from all over the Garos were brightly dressed in differing colours for each tribe. They had ten drummers apiece and dancing girls, as well as a warrior with a sword and wooden shield.

Each drum rhythm and dance was different and represented various aspects of life, the most memorable being a dance to represent *"the shooing away of flies that are perching upon rice"*.

At the festival, we noticed some other westerners who turned out to be French girls who were studying the hoolock gibbons. They introduced themselves and said that they had found our camera traps at Nokrek and apologised in case they turned up on any of the pictures. They had even written a letter to us and left it at the lodge not realising that they would meet us at the Wangala Festival.

---

* http://100drumswangala.blogspot.com/

EXPEDITION REPORT: India 2010

Later we all had dinner at a tourist lodge. The lady who ran the lodge had recently cooked for the Prime Minister of India and I could well believe it. 'Dinner' is a term that does what we ate a disservice - it was a feast of positively medieval proportions with whole chickens, huge fish, sides of pork and masses of fruit.

The following day we met another impressive witness. Nelbison Sangma was a farmer from the village of Sansasico, and he had observed a mandeburung for three days running in 2003. Nelbison was some 500 metres from the creature, and was able to look down upon it as it was on top of a smaller hill. When he first saw it, the mandeburung was standing under a tree and Nelbison told us that it was nine feet tall and covered with black hair. Whilst he watched, it moved around for an hour and then slept in a nest it had constructed by pulling down branches much like a gorilla does. The next day the creature was in the same place and appeared to be sunning itself, and this time he watched it for half an hour. On the third day he saw it again and it was wandering about and foraging.

The following day he took some other villagers to the area and showed them the nest and they noticed there was a monkey-like smell that pervaded the surroundings. They found man-like tracks 18 inches long and a huge dropping which contained banana leaf fibres and was the length of a human forearm.

We switched our attention back to Nokrek National Park. On the way, we picked up some provisions including rice, fruit and several live chickens. These I named Little Lofty, Gloria and Mr La-di-dah Gunner Graham after characters from *It Ain't Half Hot Mum*. Dave bought a pot of the rice beer we had tried at the Wangala Festival from a roadside vendor.

We stayed in a specially made tourist lodge near a village in the park, which had been constructed to look like a traditional Garo house made of wood and bamboo. During the day's exploration we came across a huge man-like track imprinted deep into the sand beside a stream, and we photographed it with frames of reference. The print was like a human track but with a couple of important differences, i.e. the heel was proportionally broader, indicating a weight-bearing heel, and the toes were more even, showing much less of curve from the big toe down to the little toe. The track sunk over an inch into the wet sand, whereas my own footprints could not reach even half that depth.

Back at the lodge we ate dinner and then sat around the fire telling stories, when I tried some of the rice wine that Dave had bought. I was to regret it later. During the night I had severe stomach pains as if someone were twisting a knife in my guts, and I felt bloated and feverish. During one of my many visits to the toilet during the night, I heard something large moving around in the darkness outside the lodge. I assumed that it was a goat or a cow that had wandered down from the village. In the morning Rudy told us that something had been pushing against the lodge door. He, however, thought that the live chickens we were keeping inside the lodge had attracted a tiger.

I was feeling worse than ever and was in a lot of pain, but I forced myself to go with the others into the jungle on the off-chance the creature would put in an appearance. Suffering from intense stomach pain and fever, I found the trek hard and at a waterhole I had to stop and rest as the others went on. Dave and Jon found another set of mandeburung tracks by a stream further into the jungle, which followed the stream and seemed fairly fresh. The creature seemed to be overturning rocks and hunting for freshwater crab and some shells were discovered with their insides sucked out. In the meantime, I managed to stagger back to the camp, and – thankfully - by the next morning, I was feeling better.

We returned to Tura, as Morgan and Tara had to leave for Delhi the next day, which we spent around Tura chasing up some leads and loose ends. We photocopied the relevant chapter from Dr Lao's book and then visited the rather shabby library to see if there was anything on the mandeburung and the sankuni in any of the books there. We turned up absolutely nothing there, so we tried to track down Albin Stone, the man who was said to have eaten the flesh of a dead, juvenile sankuni, but he was not at home.

Dipu showed us a rib bone found by his father at Balpakram in 1989. I thought it looked more bovid than primate, but we took a sample from it. Dipu also showed us a collection of hairs he found at Nokrek in 2006, which looked to me like goral (*Naemorhedus goral*) a goat-like antelope known to inhabit the area. All of these, together with the 'muntjac' sample and the bone from Siju Caves, were later sent to Lars Thomas and his team for analysis.

We met Dipu's uncle - Garfield - who, whilst fishing in 1956 or'57, came across a mandeburung print beside a stream. It was on a rock and formed by water where the creature had recently walked out of the stream and across the rocks.

Garfield also claimed to have seen the trail of a sankuni in the 1970s. The track emerged from the Garo River and went for 100 metres under a wooden bridge, destroying some of the supports, and then crossed a paddy field and entered a marsh. However, upon closer questioning Garfield told us that there were *two* tracks running parallel to each other and the ground/vegetation between them was undisturbed. This sounds very much like the tracks of some kind of all-terrain vehicle rather than a giant snake that would leave one single furrow.

All too soon, our time in the Garos was over. We had to say our goodbyes to Dipu, Rudy, Rufus and the others, before we returned to Delhi where we had a day to see the local sights. We had intended to take in the Taj Mahal but it was too far away, so we made do with the local sights such as Humayun's Tomb and the Qutb Minar. Ironically, we saw more wildlife around the city than we did in the jungle. There were flocks of ring-necked parakeets (*Psittacula krameri*), troops of rhesus macaques (*Macaca mulatta*) and Indian three-striped palm squirrels (*Funambulus palmarum*).

At the time of writing, the bone, Lars Thomas and his team are testing the bone, hair and antler samples, and the initial tests on the antlers have yielded some exciting results.

I am convinced that the mandeburung exists and that it is one in the same as the larger kind of yeti. The best model we have for this animal is a surviving form of *Gigantopethicus blacki*. As for the sankuni, its startling resemblance to the Indo-Chinese naga, the West African ninki-nanka, the Central African crested crowing cobra and many other monster serpents, convinces me that there are more to these stories than hot air.

Already there is talk of returning to the Garos in a few years time, probably to mount an expedition down into the gorge at Balpakram. Kipling's India is still alive if you look hard enough and I intend to return there.

EXPEDITION REPORT: India 2010

Jon says farewell to Richard at the exotic Barnstaple station

EXPEDITION REPORT: India 2010

TOP: Richard mounts the carriage
BOTTOM: Ladies selling their wares at a Garo village market

EXPEDITION REPORT: India 2010

A village market a few miles from Tura

EXPEDITION REPORT: India 2010

TOP: The Garos have Tibetan ancestry, hence their oriental looks
BOTTOM: The entrance to Nokrek National Park, the scene of many yeti sightings.

EXPEDITION REPORT: India 2010

TOP: Tescan, a local fruit with a unique flavour. BOTTOM: Chris samples some

EXPEDITION REPORT: India 2010

TOP: A lodge made to look like a Garo house; we had our base here whilst in Nokrek
BOTTOM: A cement tree house

EXPEDITION REPORT: India 2010

TOP: The jungle swathed Garo Hills, haunt of the Indian yeti.
BOTTOM: Removing a fallen tree from what passed for a 'road'.

EXPEDITION REPORT: India 2010

TOP: The jungle in Nokrek BOTTOM: The nest of a wild boar

EXPEDITION REPORT: India 2010

TOP: Fungus on a rotting log in the jungle
BOTTOM: The wall creating a man-made water hole

EXPEDITION REPORT: India 2010

TOP: The team at the water hole BOTTOM: The unattractive view from our hotel in Tura

EXPEDITION REPORT: India 2010

Richard in the jungle

EXPEDITION REPORT: India 2010

Dave setting up a camera trap

EXPEDITION REPORT: India 2010

Richard and Dave at the lodge in Nokrek

EXPEDITION REPORT: India 2010

TOP: View of the Garo Hills BOTTOM: The Simsang (Rice Pot) River

EXPEDITION REPORT: India 2010

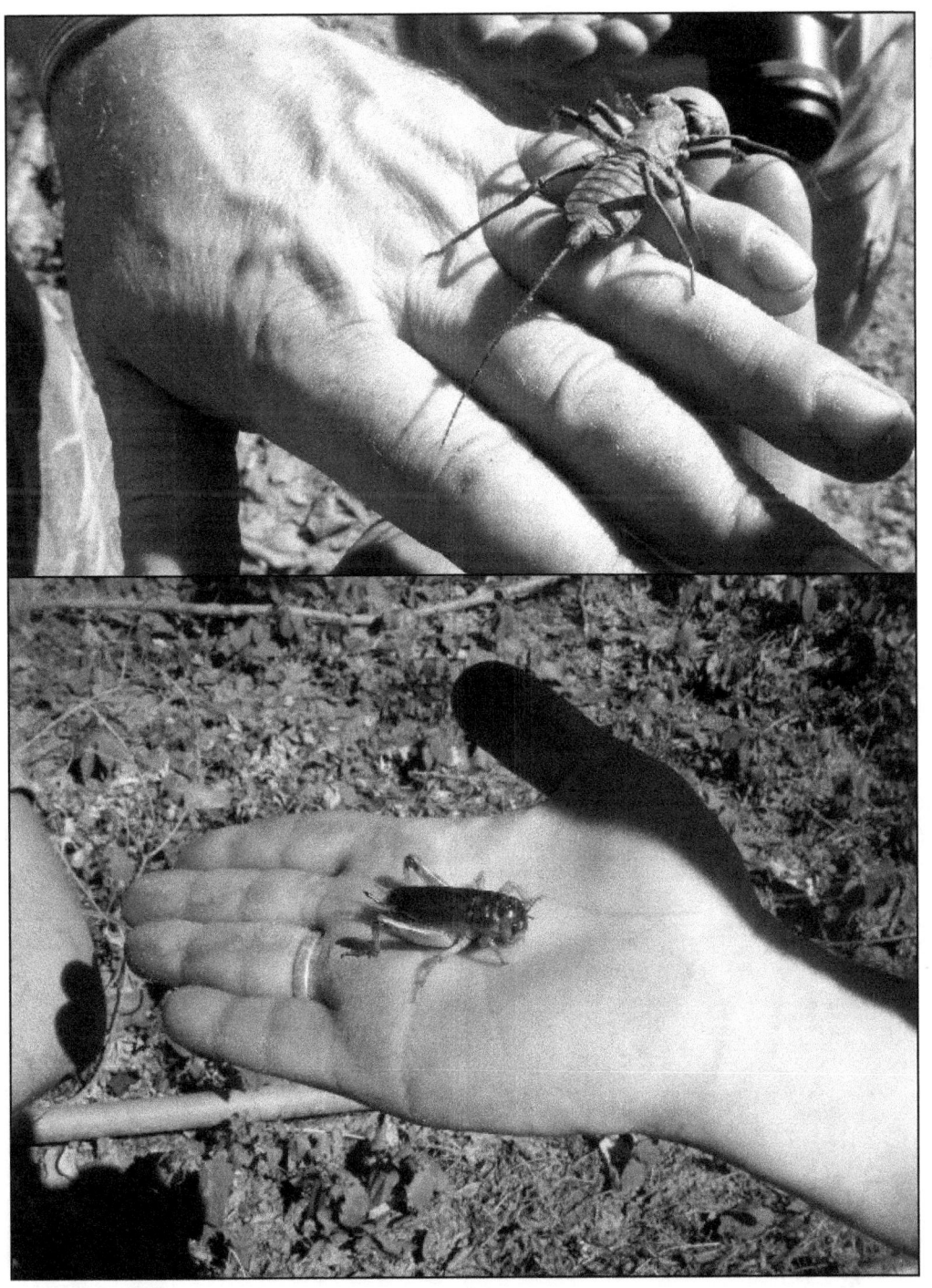

Vinegaroon whip scorpion and a huge cricket caught by Dave and Jon

EXPEDITION REPORT: India 2010

Two views of the Simsang

EXPEDITION REPORT: India 2010

TOP: Richard above the Simsang
BOTTOM: Chris, Adam and Richard in Siju Cave were a mandeburung was reported

EXPEDITION REPORT: India 2010

Chris and Dave on the spectacular bridge above the Simsang

EXPEDITION REPORT: India 2010

Richard in Siju Cave

EXPEDITION REPORT: India 2010

TOP: Stalactite in Siju Cave BOTTOM: Pintu, the guide who found the bone

EXPEDITION REPORT: India 2010

TOP: A huntsman spider in the cave BOTTOM: A cave crayfish

EXPEDITION REPORT: India 2010

TOP: A narrow mouthed frog in the cave
BOTTOM: Fulvous fruit bat (*Rousettus leschenaultia*)

EXPEDITION REPORT: India 2010

TOP: Fungus growing on bat guano
BOTTOM: Rocks in the cave, the more credulous should note the orbs.

EXPEDITION REPORT: India 2010

TOP: Spectacular stalactites at Siju BOTTOM: A mottled swamp eel caught for dinner. The mucus of this species, mixed with rice or flour is used as folk medicine for arthritis.

EXPEDITION REPORT: India 2010

Mottled swamp eel, it tasted very nice. Thanks to Shara for the identification
http://fishbase.sinica.edu.tw/images/species/Anmar_u0.jpg

EXPEDITION REPORT: India 2010

TOP: The Simsang River BOTTOM: A small landslide, doubtless blamed on the sankuni

EXPEDITION REPORT: India 2010

Limestone outcroppings that recalled the 'Cold Lairs' in Kipling's Jungle Books

EXPEDITION REPORT: India 2010

TOP: A watering hole in the jungle near Siju BOTTOM: Leopard scat

EXPEDITION REPORT: India 2010

TOP: A mantis ootheca BOTTOM: Elephant dung

EXPEDITION REPORT: India 2010

TOP: The elephant watching tower BOTTOM: A small cave in the jungle

EXPEDITION REPORT: India 2010

Richard holding, not a snake skin but the elongate honey comb of a wild bee

EXPEDITION REPORT: India 2010

TOP: Tracks of a small cat in the cave BOTTOM: The small jungle cave

EXPEDITION REPORT: India 2010

TOP: Limestone simulacrum BOTTOM: Jon examines the small cave

EXPEDITION REPORT: India 2010

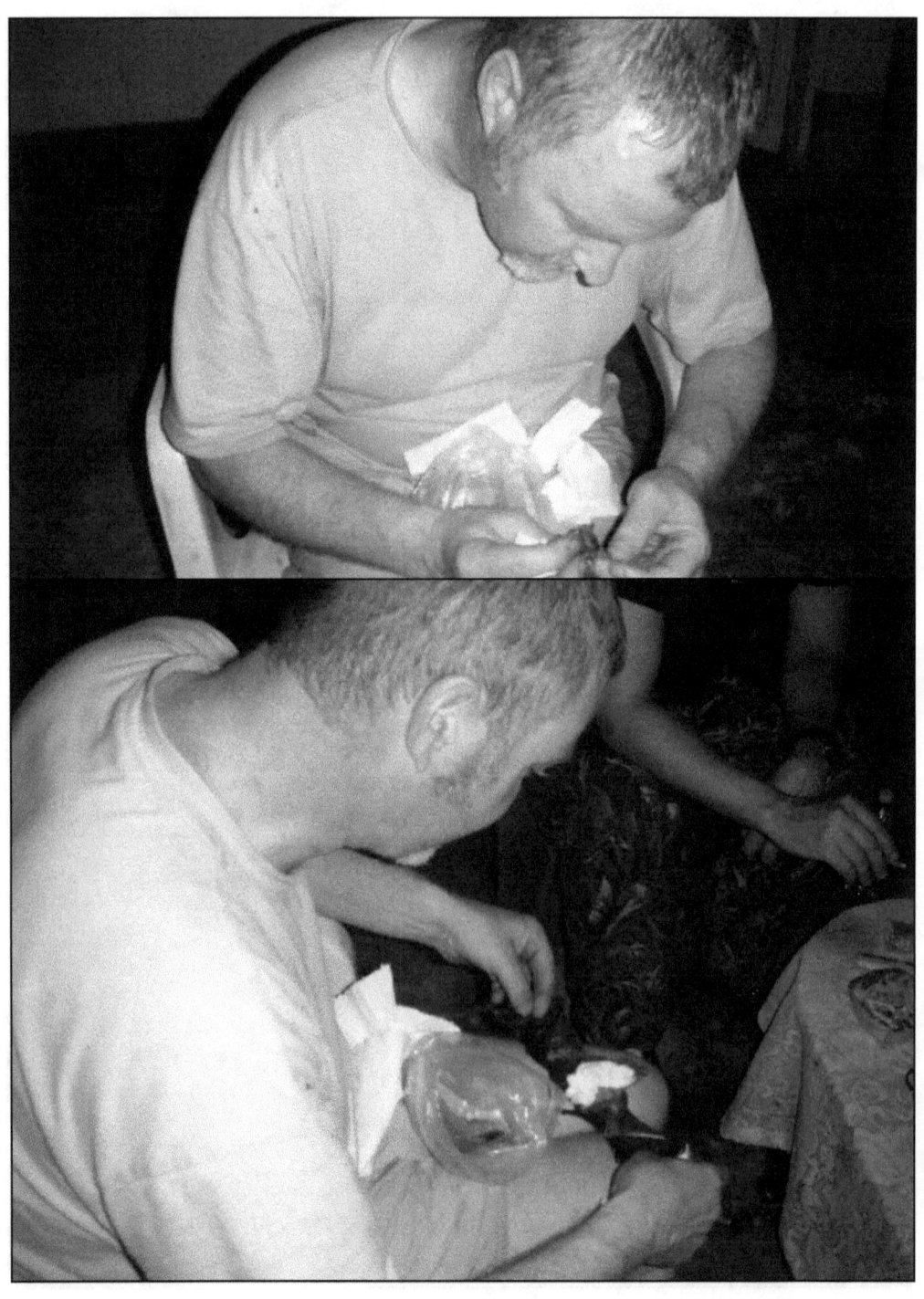

Two shots of Jon preserving dead fruit bats from Siju cave

EXPEDITION REPORT: India 2010

TOP: A tokay gecko in the Siju lodge BOTTOM: The kitchen with the oppressive atmosphere

EXPEDITION REPORT: India 2010

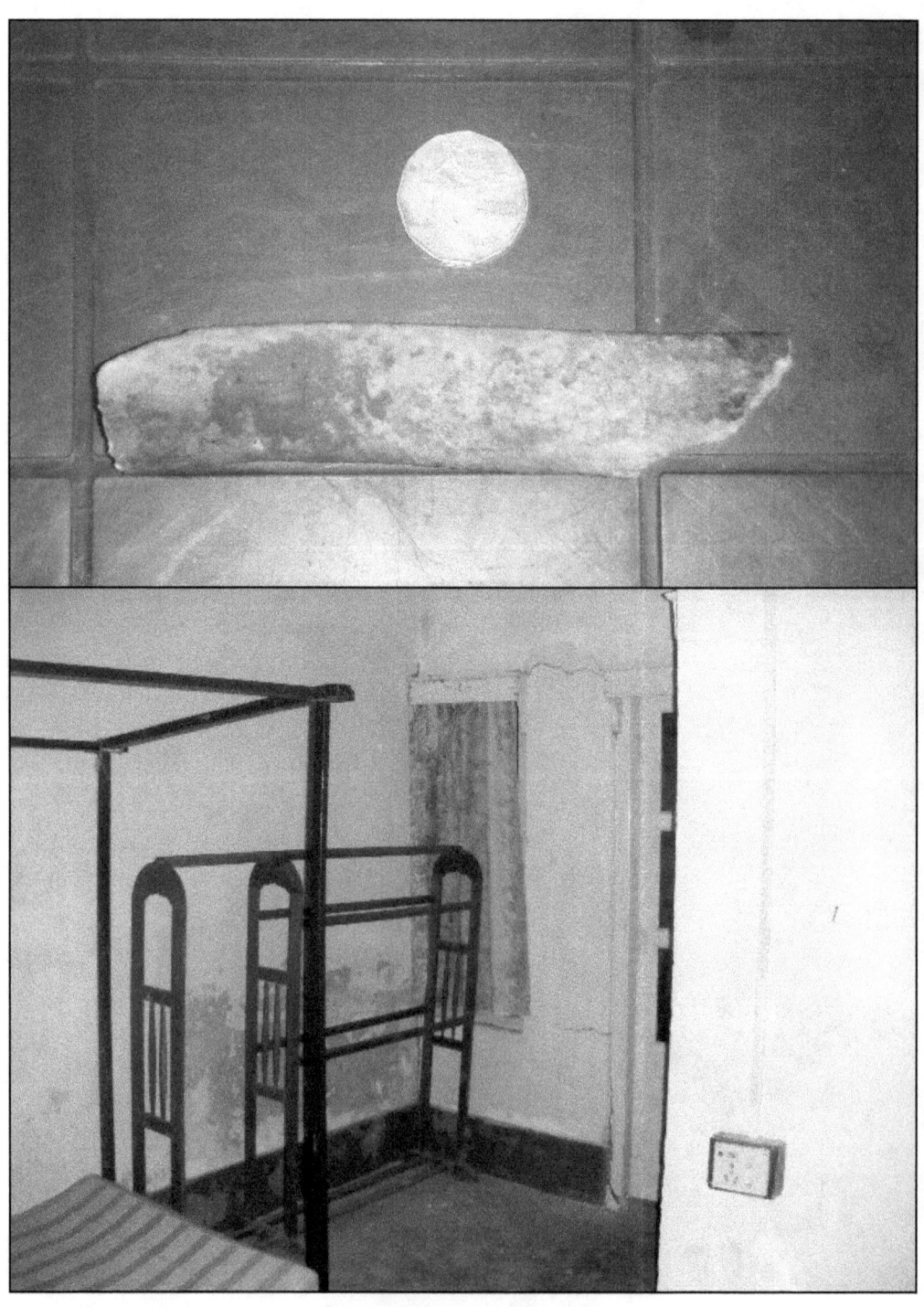

TOP: The tibia fragment found by Pintu at Siju Cave
BOTTOM: The haunted bedroom at Siju lodge

EXPEDITION REPORT: India 2010

Gentar, head man of Siju - he says he encountered a mandeburung in the cave

EXPEDITION REPORT: India 2010

TOP: Tarantula fishing BOTTOM: View of the Simsang from Bagimara lodge

EXPEDITION REPORT: India 2010

TOP: Dead white lipped viper in the jungle beyond Imangri BOTTOM: A stream in the jungle

EXPEDITION REPORT: India 2010

TOP: Simulacrum of yeti footprint near the jungle stream beyond Imangri
BOTTOM: Another view of the stream

EXPEDITION REPORT: India 2010

TOP: A canoe on the Simsang BOTTOM: A swarm of yellow butterflies by the stream

EXPEDITION REPORT: India 2010

Shireng R Marak who hid all night in a cave from the mandeburung

EXPEDITION REPORT: India 2010

Pet hornbill at Imangri

EXPEDITION REPORT: India 2010

TOP: Dave and Chris in Imangri village
BOTTOM: Rudy and Adam with yeti witness the shaman Neka Marak

EXPEDITION REPORT: India 2010

TOP: Silkmoth caterpillar
BOTTOM: The jungle hills stretching way to the borders of Bangladesh

EXPEDITION REPORT: India 2010

TOP: The burning of stubble at Balpakram BOTTOM: Cactus-like plant at Balpakram

EXPEDITION REPORT: India 2010

TOP: Richard with Balpakram gorge in the background BOTTOM: The undisturbed gorge

EXPEDITION REPORT: India 2010

TOP: Chris Clark by the gorge BOTTOM: The team L-R Richard, Adam, Dave, Jon, Chris.

EXPEDITION REPORT: India 2010

The cooled lava rock formation at Balpakram recalls the Irish Giant's Causeway (above)

EXPEDITION REPORT: India 2010

TOP: A western style coat of arms with Thai writing at the hotel in Tura, we never did find out what it said BOTTOM: Richard at Bagimara reading Kipling's classic.

EXPEDITION REPORT: India 2010

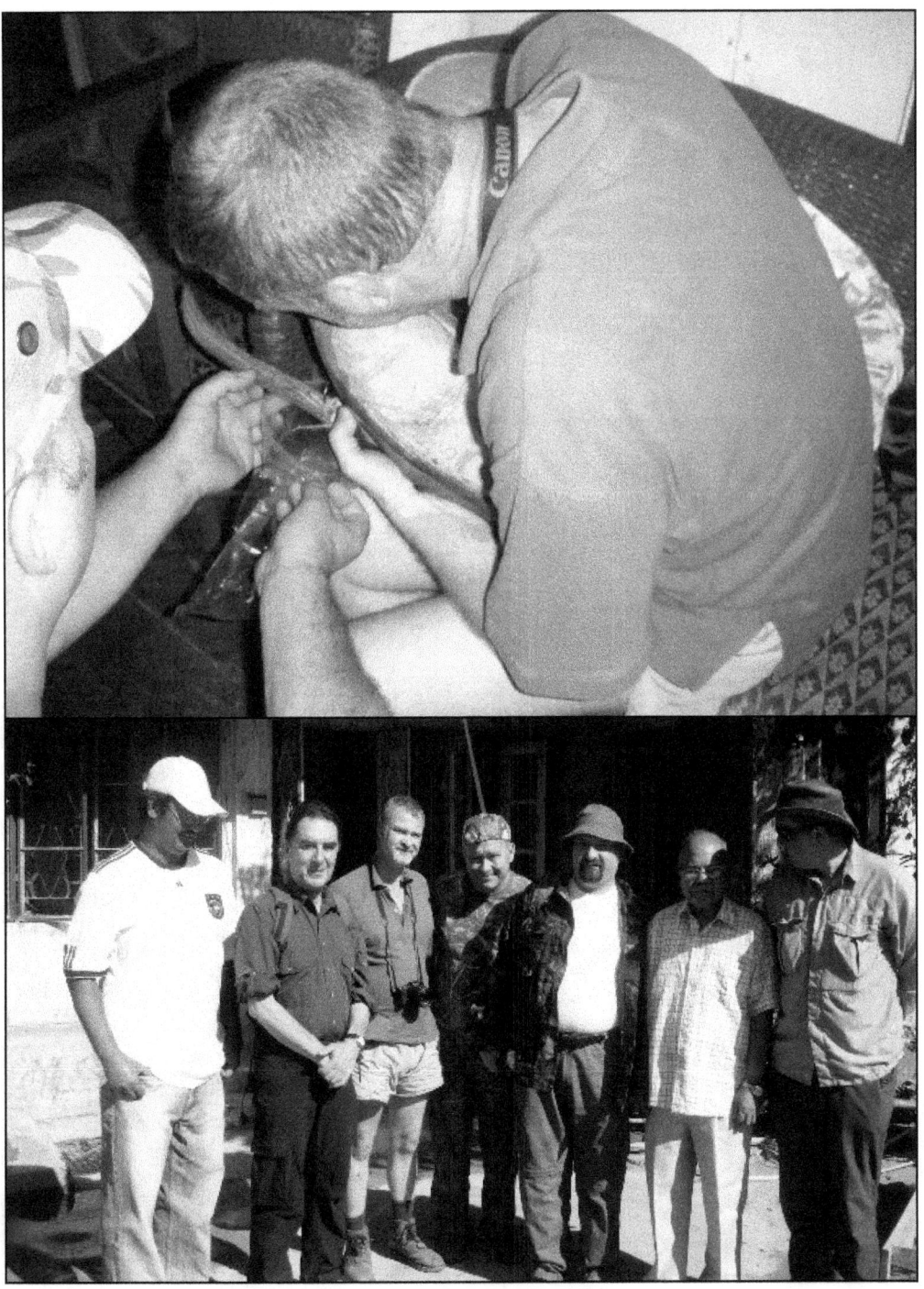

TOP: Jon and Dave take samples from antlers of what we thought might be a new species of deer. BOTTOM: The team with Dipu Marak, and Dr Milton Sasama

EXPEDITION REPORT: India 2010

The Garo naturalist and author Llewellyn Marak

EXPEDITION REPORT: India 2010

Dr Lao with the skin of a red panda shot locally in the 1960s.

EXPEDITION REPORT: India 2010

Burmese python at the Tura zoo

EXPEDITION REPORT: India 2010

Sloth bear in inadequate enclosure at Tura zoo

EXPEDITION REPORT: India 2010

TOP: Indian soft shelled turtle kept as a pet in Tura
BOTTOM: Dave hunting snakes by the roadside

EXPEDITION REPORT: India 2010

Waterfall on the outskirts of Tura

EXPEDITION REPORT: India 2010

Kingston, who heard the yeti's cry and found its tracks on Tura Peak

EXPEDITION REPORT: India 2010

Nicholas Sama who saw a preserved mandeburung arm at a bush meat market in the '60s

EXPEDITION REPORT: India 2010

The face of the future, android courtesans?

EXPEDITION REPORT: India 2010

Drawing of a female mandeburung with infant made by Dave under the instruction of witness Teng Sangma

EXPEDITION REPORT: India 2010

Teng Sangma who saw a female yeti suckling an infant whilst it fed on bamboo.

EXPEDITION REPORT: India 2010

Teng shows us the area of the jungle were he saw the creature

EXPEDITION REPORT: India 2010

The stand of bamboo on which the creature was feeding

EXPEDITION REPORT: India 2010

Jon examines footprints, they are clearly human

EXPEDITION REPORT: India 2010

Garo tribal warrior at the Wangala Festival

EXPEDITION REPORT: India 2010

Teams of drummers, dancers and warriors perform in the inter tribal contest of Wangala or the Hundred Drums Festival

EXPEDITION REPORT: India 2010

Drummers and warrior at the Wangala Festival

EXPEDITION REPORT: India 2010

TOP: Spectacular feast held at a new tourist lodge, just for us!
BOTTOM: Adam tries his hand at a local drum

EXPEDITION REPORT: India 2010

TOP: Pots of rice beer for sale on the road side BOTTOM: This lady cooked our feast; she also recently cooked for the Indian Prime Minister

EXPEDITION REPORT: India 2010

Nelbison Sangma who saw the yeti on three subsequent days

EXPEDITION REPORT: India 2010

Another man from Sansasico village who saw huge footprints

EXPEDITION REPORT: India 2010

A jungle glade in the forest where Nelbison had his encounter

EXPEDITION REPORT: India 2010

TOP: Giant pond skaters on a stream BOTTOM: Stream where we found the yeti's spoor

EXPEDITION REPORT: India 2010

TOP: Possible yeti track in damp soil by the stream. 12 inches long and 2 inches deep
BOTTOM: Richard's hand next to the track for comparison.

EXPEDITION REPORT: India 2010

TOP: Richard's weight could only make a slight impression on the tightly packed soil. The track however sunk in 2 inches! BOTTOM: Richard's size nines for comparison

EXPEDITION REPORT: India 2010

TOP: Nelbison smoking a traditional bamboo pipe BOTTOM: Nelbison's friendly dog

EXPEDITION REPORT: India 2010

Adam and Chris taking tea in the lodge at Nokrek / Adam looks on as Dave opens a road side rice beer pot

EXPEDITION REPORT: India 2010

TOP: Chris sucks rice beer through a bamboo straw
BOTTOM: Richard drinks the rice beer that violently disagreed with him!

EXPEDITION REPORT: India 2010

TOP: Dave examines a camera trap laid a fortnight earlier at Nokrek
BOTTOM: Fossilized nautilus shell found by Dipu

EXPEDITION REPORT: India 2010

Chris examines a camera trap

EXPEDITION REPORT: India 2010

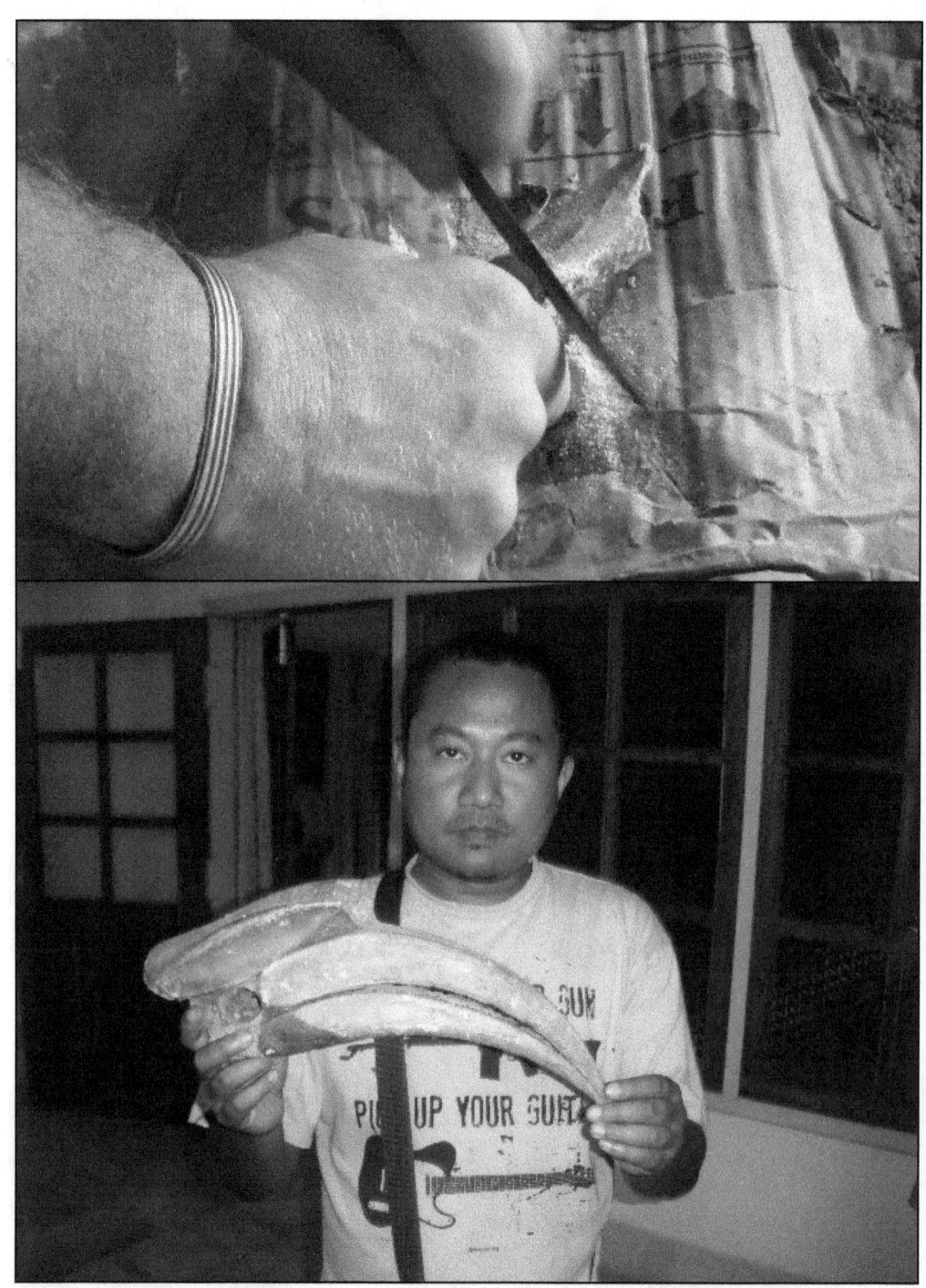

TOP: Taking a sample of the tibia found in Siju Cave
BOTTOM: Rudy our guide and Garo folklore expert with the skull of an Indian hornbill

EXPEDITION REPORT: India 2010

TOP: Spectacular root system of a banyan tree in Tura
BOTTOM: One of the thousands of five lined palm squirrels in Delhi

EXPEDITION REPORT: India 2010

TOP: A nest of huge wasps in Delhi
BOTTOM: Adam at India Gate, in front of the world's crappiest ice cream van.

EXPEDITION REPORT: India 2010

The qutub minar in Delhi

EXPEDITION REPORT: India 2010

A ceremonial gate in the grounds of the qutub minar in Delhi

EXPEDITION REPORT: India 2010

Two faces of Delhi - ancient and modern

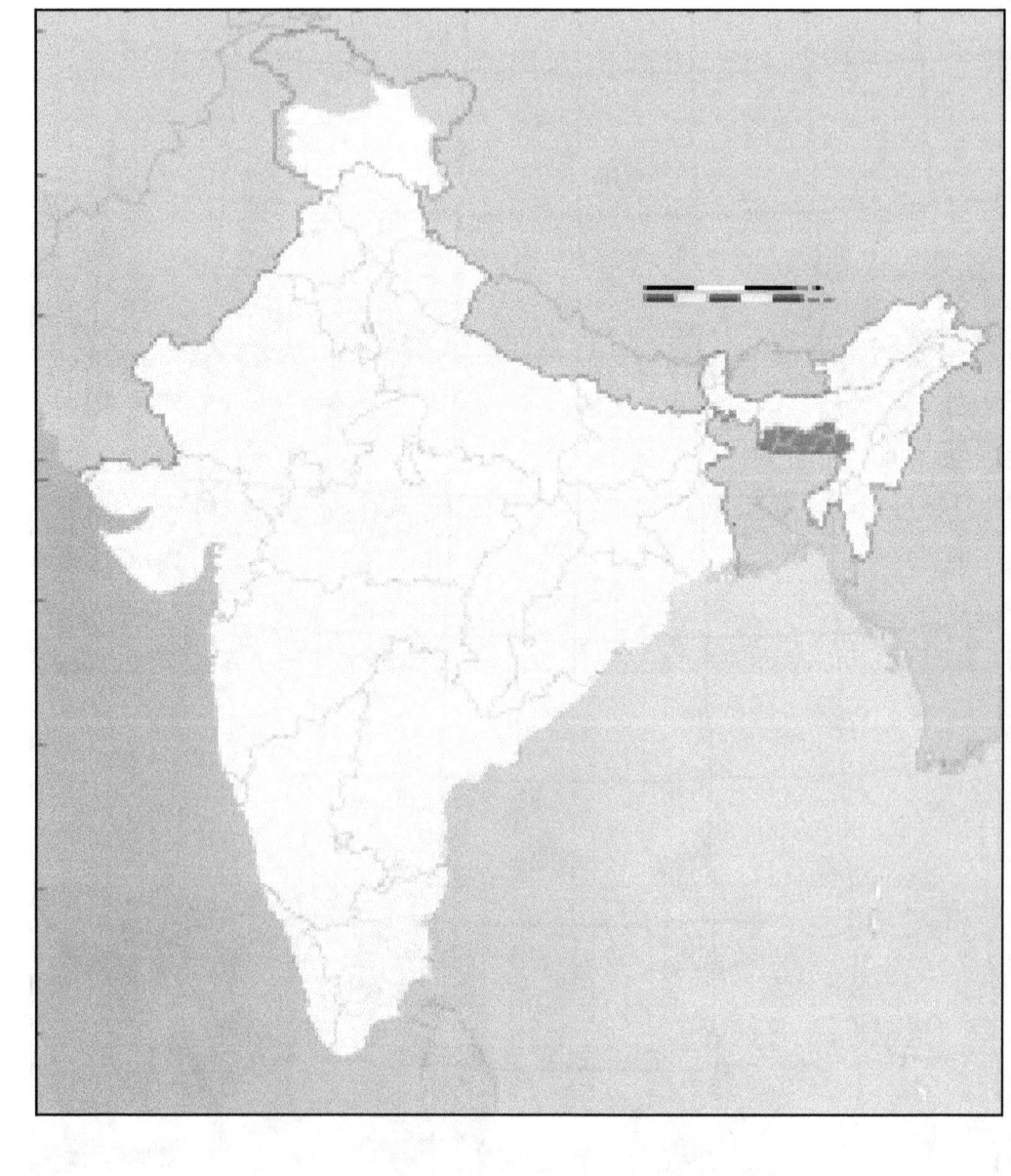

MY CLOSE ENCOUNTER WITH THE MANDEBRING

RUDY SANGMA

Some time or other, we have all heard news about sightings of gigantic, hairy, ape-like, bipedal creatures that resemble no ape known to science. Here in the Garo Hills, we also have such a creature. It is called the *Mandebring* (mandebrɨŋ). The word *Mandebring* is derived from two Garo words, *mande* meaning "man" and *bring* meaning "forest" or "jungle" – forest man or jungle man. Other spellings of this that can be found on the internet are *Mandeburung* and *Mandeburing*. That such a creature, unknown to science, exists has neither been proved nor disproved beyond the shadow of a doubt.

I have been fascinated with the paranormal as long as I can remember, with books like Reader's Digest's *Strange Stories, Amazing Facts* and Karl Shuker's *The Unexplained* keeping me interested for hours on end. So it was natural that I would look around for paranormal anomalies in and around my immediate environment - and my tribal culture provided me with a treasure trove. We have, in the folklore of my people, the Garos, quite a large number of paranormal anomalies, among them cryptids, and among the cryptids is the *Mandebring*. I have heard of tales of the *Mandebring* ever since I was a child, but in my memory, the first ever widespread news of it was in 1997 when huge human-like footprints were found in the banks of a stream called Andol Chiring in a place called Silkigri. My friend Dipu N. Marak has been following it ever since and over the years has collected material from different sightings all over the Garo Hills.

My induction into the quest for *Mandeburing* was in the year 2008 when BBC reporter Alistair Lawson came to Tura to do a story on the creature. I accompanied him as his guide and interpreter and since he was alone, had to double up as his cameraman. We went deep into the jungle near Nokrek National Park with Nelbison M. Sangma, a villager of Sasatgre village, who had observed the creature for three days in 2002. He took us to the place where he saw the creature make a nest, sun itself and run around. We had to trek about two to three hours through impenetrable jungle, steep terrain and cross a river to reach the place. The place was on top of a hillock and overgrown with thorny vines and thatch grass, but not many trees because, according to the eyewitness, the creature had practically cleared it during its stay there. On our return, we had to climb down a steep slope, which was pretty hard since I had a video camera (a Sony VX1000) on one hand and had to hold on to plants and rocks to maintain my balance. Then we had to cross the river, walk along it for some distance and climb up another steep slope along a dry stream bed. It was the toughest trek I had ever had in my life and I was so tired. After Mr. Lawson left, my interest in this creature grew and I started garnering whatever bits and pieces of information I could find.

The biggest break for us came in the spring of 2009 when there was a sighting in a village called Rongbakgre about 15 kms from town and driveable all the way. All the previous sightings had been in far off places which could only be reached after a long trek from the nearest road. Though the village is very close, the news of the sighting reached us only after two weeks. The first sighting was by a carpenter who was working in the village. He had gone into the forest by the village early in the morning to answer the call of nature when he saw the creature. He was so spooked by it that he immediately ran off to his village, which is about 70 kms from there, and fell sick.

Only after he reached his village did he call his employer and inform him that he had seen a demon. According to the employer, the carpenter told him that the creature had hair all over, the hair from its head reaching the waist and it was suckling an infant. As soon as we heard the news, we reached the place, took along some villagers and went into the forest, a bamboo reserve kept by the village. Though none of them had seen the creature, they showed us signs of the creature having been there: like the barks of trees being worn due to the creature rubbing against it, the ground cleared of leaves, etc. There is a tree locally called *te·wek* that bears fruit, the size of olives, in bunches, not only in the branches, but also from the trunk, right from ground level, to the top. The fruit is usually abundant in such trees, and not many animals eat it. That day, we saw one from which every single fruit, numbering in the thousands, had been plucked with no sign of a single fruit fallen on the ground. The tree was about five to six feet in diameter and the branches were pretty high up. We also observed some claw marks at the bottom of the trunk. Now, the only creatures living in that forest were small game, jungle fowl and some monkeys. Amazingly, even the monkeys seemed to be absent that day. According to the villagers, they were following the creature, although I really don't know where they got the idea.

A couple of days later, at around 8 p.m., Dipu called and said that there had been another sighting in the village. This time it was only a few hours old. So we went there immediately and interviewed the witnesses. The witnesses were a couple of men from the village who had gone to the forest to hunt for fowl and small game. According to their account, they had gone hunting around 4 p.m. and were walking along a stream. As they turned a bend in the stream, they saw the creature sitting with its back towards them on the bank, which was at an elevation of about four to five feet above the stream. They also saw an infant with it. They were terrified by it and ran off as silently and as fast as they could, and warned the villagers from going into the forest. We were also informed by someone who knew our interest in this subject.

The next day, we organised a search party with some people from Tura and some from the village. The total number of the search party was over a hundred men. We split into groups and combed the forest. We did see signs of activity, but though we searched the whole day, we could not find anything. We did see places where it slipped in the loose soil, but we could not find a single clear footprint. The ground was covered in leaves and maybe the creature was trying to hide its tracks. When my party got to the place where the two men had seen the crea-

EXPEDITION REPORT: India 2010

ture, we did see signs of activity there and also some droppings on a rock in the stream bed. We were not really interested in it, though in hindsight, we let a very good piece of evidence go to waste. The droppings looked as if it could have been made by the infant and looked like seeds from the fruit mentioned earlier.

While the other teams slowly drifted back from the forest, my team and I kept looking. We went down one stream and came up another until we reached a small gorge, 90 degrees on one side, and a steep 75 degree slope on the other. It was about 2 p.m. in the afternoon when we reached there and took a rest. We were just about to give up when we noticed some activity overhead. We looked up and we saw some monkeys, the first sign on any fauna we had had the whole day. I remembered villagers telling me that the monkeys followed the creature. We decided to investigate and started to climb up the slope. Halfway up the slope, I got a very pungent smell. We reached the top and began our search once again and came to a clearing. The clearing was in the middle of the bamboo forest, which had leaves all over the forest floor, but here, all the leaves had been swept clean. Although close to human habitation, there were no signs of human activity in that area. The bamboo growth was so dense that you could hardly make out what was ten feet in front of you. In the clearing was a small tree whose bark was worn as if something had rubbed against it. We moved away from the clearing and waited for the better part of an hour. The monkeys were still there but they were very quiet. There was no more activity in that area and we returned from there because we had been searching since morning and were feeling quite hungry and tired. We made our way through the thick bamboo and came out into an area which had been cleared for cultivation. I reported my findings to Dipu and some of us decided to return that very evening to the same place.

That evening, we returned with torch lights. It was still light when we reached the cultivated area. Since it would soon be dark, we decided against going back into the bamboo forest, to the clearing. Instead, we decided to climb down the gorge along the stream. As we were coming down towards the stream, atop a leafless tree, we saw two big monkeys looking around and when they saw us, they screamed and made as if to climb down. When we stopped, they too stopped and looked at us, wary of our movements. As we sat in one place looking at the monkeys, we heard some noises from the gorge. We heard loud grunts, grunts that could only have been made by a large animal and some chattering sounds. There was also the sound of water splashing. Since the sounds seemed to be coming upstream, we waited with bated breath at the chance to see this fabled creature. As we were waiting thus, all of a sudden, we heard the sound of people singing and coming upstream. We were to learn later that they were people from town that had come to try their luck at getting a glimpse of the creature and as it was getting dark, they were scared and so came along singing loudly. As soon as the monkeys heard the singing coming from downstream, they screamed and immediately climbed down the tree. Then there was such a commotion in the undergrowth by the banks of the stream. We saw a whole clump of bamboo and trees shaking with the strength that only a big and strong creature could possess. After that everything was quiet. The people who were singing came upstream and crossed us. By now it was already dark. We waited for sometime afterwards but there was no more activity there. There was no sign of the monkeys, too. It was as if those monkeys were posted as sentries for the creature. A friend of mine, an unbeliever said that he had to believe that the way the bamboo and the trees were moving, it had to be some sort of a gigantic creature.

The next day, we went back around the same time as the previous day. We decided to go down to the gorge. There we saw broken trees, trees whose trunk would've been about nine to ten inches in circumference. Then we climbed to the other side and found the reason why we could not make out the creature's departure from the area the other night. The whole area was cleared and had been converted into a betel nut plantation. We wandered around the forest for a couple of hours but could not see signs of the creature nor the monkeys. I will always remember the day I had a close encounter with the *Mandebring* because that day, my belief in the existence of the creature became concrete.

EXPEDITION REPORT
ADAM DAVIES

Expedition start up:
After the drama of Sumatra had settled, I began to plan where we might go next.

I had heard that there was a team of researchers in India that, for several years, had been coming up with some evidence of a large bipedal primate there; indeed, I had even advised the producers of *Monsterquest* that it would be a good story to check out. Subsequently, they sent a team on a short recce to the Garo Hills for their 'Curse of the Monkey Man' episode -though whether it was down to my instigation or not, I cannot say.

However, it was when I met up with Ian Redmond in the Himalayas that my interest was really fired. Ian was part of my team which completed a show for *Monsterquest* entitled 'The Abominable Snowman'. Ian has an OBE, is the U.N. Ambassador for Gorillas, and is obviously a knowledgeable guy.

I had been at altitude in the mountains, whilst Ian had remained at a lower level in the forested area, looking for evidence of the Yeti. Over our rendezvous in the company of a whisky, we discussed the feasibility of other cryptids.

Ian had been involved in the testing process of some alleged Mandeburung hair, which sadly turned out to be from a mountain goat called the goral. That said, he was impressed with the research coming out from an energetic researcher called Dipu Marak, along with the weight of evidence he produced. So, fired by his enthusiasm, I decided to contact Dipu and do some research myself.

Whilst beginning my research, I was struck by the fact that the jungle in which the creature was said to roam was infrequently traversed. Dipu was also able to confirm that he knew eye-witnesses, and was able to verify locations to where it was feasible to travel. Thus, I was satisfied that it would be an excellent place to take a CFZ team, and began to plan the expedition in March 2009.

Going to a place where little research has been conducted is not like going to an area like Sumatra, where I have been many times before. It requires a lot of planning, and some degree of risk taking - it was always in my mind that I was asking people to take personal and financial risks on an adventure, which could turn out to be a dud.

However, over the months Dipu and I exchanged hundreds of emails about everything, from kit, to terrain, to equipment, and I found him to be a meticulous planner and organiser. One thing Dipu was keen to warn us about was the potential terrorist activity.

The indigenous peoples of the Garos have their own distinct culture and traditions. They are Asiatic in appearance, looking rather like their neighbours the Butanese. They are also mainly Christian (with some animists), unlike the vast majority of Indians who are of course, either Hindu or Muslim.

Dipu warned us that a small minority of members of the Garo tribe supported the terrorists. One group, the larger one as it turned out, had come to some sort of ceasefire with the government, whilst the smaller one, the Garo National Liberation Front, had not. They had apparently been active in the area where we were due to go – they had been involved in the kidnapping of an official, and had attacked and recently burnt out a truck on the highway (which we were to see). Of course, I discussed all this with the others. However, they were - and are - a committed bunch, and nobody backed out. So, in November we were set to go.

Finding the prints at Nokrek:
As we moved towards the back of the waterhole, I found what looked to be Mandeburung prints by the side of a small stream. They were deep, making an impression of about an inch into the ground, and were about twelve inches in length, though it was difficult to be exact as the heel imprint did not have clear definition. The toes were clearly visible as well. I was excited by the prospect, and beckoned Tara and Morgan over saying: "Hopefully, we will see evidence of the creature eating or foraging". Just ten feet up the trail, I did indeed find a boulder with a tidemark where it had been recently overturned. I felt an adrenaline surge, like the previous year in Sumatra, when Dave and I had been on the trail of the Orang-Pendek.

"We may find some debris from it eating," I said. At that point, Rudy - who had moved slightly up the trail from me - beckoned me over to inform me that he had found just that. We followed the trail from there until we lost it, as the stream took a sharp angle over the hillside and it dissipated to nothing. Sadly, there was no sign of the Mandeburung, but it had definitely been there within the last day or so.

What do I think it is?
Having visited the Garo Hills, I have no doubt that the eco-system could support such a creature as the Mandeburung. The forests with their abundant flora and fauna are largely unsurveyed. For example, many of the areas the team operated in had only had 30% surveys, as admitted by a Forest Official we interviewed.

I am of the opinion that the Mandeburung is a relic Gigantopithicus or some derivative of it. In synopsis, it appears to be 9 to 10 feet tall, bipedal, and an opportunistic feeder, its diet consisting of anything from bamboo to local fruits and freshwater crab. Like other large apes, it seems to build ground dwelling nests, and like the gorilla may indulge in mock charges to scare humans away. Understandably so. In the remote jungle it frequents, the only humans it is likely to encounter are hunters.

THE SEARCH FOR THE MANDEBURUNG
JONATHAN MCGOWAN

The thought of actually taking part in an expedition to hunt for one of cryptozoology's most enigmatic creatures was too attractive. I had always wanted to do just that. I used to dream of going to British Columbia or Alaska and tracking Sasquatch through the mountainous forests. When it comes down to cryptids, I had - and still do have - more interest in the hominid and ape species than any other. This may be because there is more circumstantial evidence for these species, and also because the animals are possibly related to us, which makes it easier to relate to in a way, as well as making it more mysterious as to the origins of our ancestry.

It seems more real, and more likely - or more feasible - that ape-man could still be surviving today as we know they once existed in their many forms. That does not mean that I necessarily believe that the animals seen by modern man on all continents today, or in the near past, are ancestors of man at all; no. They could be something completely different - not hominids at all - but it would seem more likely that they are, going by the many reports from around the world.

As there have been many hundreds of reports of these animals, and possibly thousands of sightings not yet reported, one has to conclude that something exists. These credible witnesses have come from all walks of life, and are - more often than not - naturalists or zoologists, teachers, servicemen and women, doctors, etc.

It is easy for the sceptic to brush it all aside and explain it away as: misidentification, people living in hope, bears, other animals, hoaxes or whatever, but when one actually speaks to one of these confused witnesses, it seems there is something more to it. Some things are very lacking in our selfish world these days, for example: proper investigation, involvement, debate, or respect for people.

Of all the strange goings on in the world and the many debatable – or controversial – subjects discussed, it is worth delving into and researching. After donating time into actually looking into the subject, some sceptics are often won over, and then one may even realise that there are many sides to a story. For every myth, fairy tale, religious writing, or just about anything that seems unbelievable, there is at least one grain of truth in the matter. There are not thousands of people who claim to see Santa Claus, or the Bogey Man, yet we expect our children to believe in it. Scepticism is - in my opinion - a psychological dysfunction.

It is worth noting that many of the known species of animals today were once thought to be just tales from idiots and people with over-imaginative minds, until - that is - science eventually sought them out. Scientists tend to build barriers and refuse to cross them, from fear of ridicule, loss of a job, peer pressure, or just uncaring selfishness. Topics steeped in controversy are ignored. Scientists do not want to waste time or money on a worthless exercise. I may ask the question, why are scientists trying to find out the answers to the universe, or the answer to everything, like the big bang? Does it really concern our immediate survival? No, of course not, so let's put things into perspective.

We are simply talking about an ape. Why the big deal?

Perhaps it is too near to home. Perhaps it is too obvious, too sensitive. It concerns a possible human relative, a possible evolutionary link from prehistoric ape, to modern man. That is why science is not too keen to delving into it for obvious reasons. In the past two hundred years various theories have been put forward as to how we evolved from certain ape-like ancestors, none of which can be proved. However, science is happy to let millions of children and adults alike be conditioned into thinking that it is right. This problem runs through society like a derailed freight train. Many subjects are wrong and as man gains more knowledge of various truths, things will change over the course of time, but will it all? How much truth is not told to us? Or are we kept in the dark with regard to certain truths or new revelations regarding science? I believe so. I actually believe that many lies are told to us regarding many subjects.

It seems ironic then, that large human-like apes cannot be known to science, yet reports go back hundreds of years! How can this be? So much money and resources are put into researching other species, yet something that could be as important as this is totally ignored. That is human nature. It must change if we are to improve as a race, but that is perhaps wishful thinking. I could rephrase that to say that it is worth us knowing the truth before we destroy ourselves. The religious fraternity would not allow any truths to get in the way of traditional thinking and teaching, especially if it could throw out the whole belief system. Maybe that is one reason as to why no research has been carried out, especially in the USA.

The animal itself may be one basic type. The animals seen all around Northern and Central Asia are more likely to be the same, but the animal seen on large islands maybe something different.

It makes sense to presume that the animal seen around the whole of the Himalayan region, and other parts of Eastern Europe or Russia is of a similar species. It may also be the same as the animal seen in North America and Canada. Just where the Australian animal comes in though, I cannot say.

The animal is obviously very intelligent, for many reasons. It seems to live in the most remote areas, away from built up areas of human habitation, and sometimes the most remote areas on earth.

In some cases, these areas seem to be pretty devoid of life in the form of vegetation, which means that the animals would have to travel vast distances not just to find food, but also perhaps even others of their own kind if they are very rare. I do not believe that they live in the high peaks of the Himalayas because there would not be much food for them, even if they are omnivorous. However, I think that they could traverse huge distances with their amazing strength and stamina taking the easier passes over the high mountains. I believe that this is when people on mountain treks sometimes see their footprints. A large ape like this would not be able to live in snow conditions; there are only a few species of primates that can tolerate cold conditions and they have adapted certain life styles to cope with high altitude seasonal winters. Large herbivorous apes would not survive in cold conditions because they would need to eat a lot of greens, as vegetation is not as high in calories, so the animals would not be able to keep warm, especially if they are moving around a lot. The animal must live mainly in warm areas throughout most of the year eating much vegetation and animal life. According to eyewitness reports, the animal stalks or runs down bovids. To do that it must be rather intelligent, similar to a predator like a cat. Animals that stalk need a lot of brainpower. The animal must be very clever and know where to find food and shelter. It must be more intelligent than orangutans, gorillas or even chimpanzees. And to cap it all, the animal seems to be nocturnal. It may be that man has forced it to be, and that means that it must have been hunted at some time. It is not likely that a nocturnal great ape has evolved so to be. It will not have excellent night vision, but many reports of it having red eyes are confusing if they are to be believed, but maybe the eye shine is reddish like many animals that have moderately good night vision.

People have witnessed the Bigfoot of the Americas swimming very well. This is odd, as most apes cannot swim at all, and only some monkeys do. If this animal can swim well – the breast-stroke as well as diving under water - then it is more likely to be related to man than any other primate.

It is an oddity, if such an animal existed, that have we not found it. But, we must consider that it does live in the most remote parts of the world and could be rare. However, taking all the hundreds of accounts into consideration, it does appear to have been seen. Something must exist. And I believe that it does.

When the chance came up to investigate and search for the Yeti in the north-eastern Indian state of Meghalaya, I was very interested. The area is not known to many Westerners, and it is not typical Yeti country as we know it, yet in a way, it seems more logical that such an animal would live in that area. It seems as though it is its ancestral home - the warm jungles and ever-green forests. It is just a stone's throw away from Bhutan, China and Burma. Reports go back a long way from the Garo people, the native people that settled here from Tibet and Burma hundreds of years ago. The animal would have been more plentiful then, but what *is* plentiful when considering this kind of animal? We do not know. Maybe family groups consist of five to ten members, or maybe they are solitary? There have been sightings of several together. I think that the animal would be solitary, and that the females keep their young until they reach sexual maturity as most large apes do. I also think that family members often visit each other, and males roam large areas, often interacting with their own family groups. I do not think that such a large animal would be communal in the way gorillas are, but more like the orangutan.

The quest to find the large ape-man is running out. I do believe that this animal, or animals, is on the brink of extinction, and it is only for its cunning and intelligence that it has survived for so long under man's modern ways. I wonder what the poor creatures think about, when they are searching for members of their own kind.

Many people do not take the subject seriously, and even within the cryptozoological community there are people who see these mysteries as something not to be found - something to always keep at arm's length.

That is how many people would want it to stay.

The serious zoologist or biologist who actually believe in its existence (and they are few and far between) should be sponsored to do a proper search, and then take years so to do. That is what needs to happen.

I would never have expected that sometime in my life, I would be in the streambed of a jungle in Meghalaya, looking at the footprints of an animal local people call the Mande-Burung.
It actually happened.

Before the expedition, I had not even heard of the north-eastern Indian state of Meghalaya, but I had, of course, heard of Assam. It had never occurred to me to ponder over a map and imagine the areas in the world where this kind of animal existed. However, once I had arranged to go on the expedition I closely looked at maps of Asia, and especially India, Myanmar, China and Bhutan. I realised that the animal needed to be as far away from humans as possible, without living high in the mountains of the Himalayas. There also needed to be warmth and ample food and cover. There were many places in the areas that had most of these components, but some areas seemed more likely than did others. Meghalaya was one that stuck out, as well as the areas of Nagaland and other areas between India, Burma and Southern China. I looked at the other wildlife - the biodiversity of certain regions around. Meghalaya seemed to be populated with humans, but it was not until I got there that I realised that, although it is heavily populated in many areas, there are many areas of wilderness and places where such animals could possibly survive, even in the presence of man at some point. If all their needs were there, together with a small amount of people, then perhaps they would put up with the risk of one or two run-ins with man. The ability to hide must be good, and a total change in behaviour must be made in order to survive gun-happy hunters. There are many areas where the Garo people do not go, even in daylight. There are a few national parks where hunting is banned, although there is always going to be poaching, and most of this will be at night. The Mande-Burung would have to be on its guard all the time, with a true fear of man.

The Garo hills are low, but plentiful. The landscape undulates, and range after range traverse the region covered in jungle. It is a real mixture of evergreen forest - mainly young - but also old forest in the national parks. Many of the larger old trees no longer exist, but there is little evidence of large scale destruction. There is currently no large commercial logging going on in this area, and what slopes are cut, are quickly covered in growth in the form of creepers and vines, palms and bananas. The undergrowth appears thicker than the understory of the old forests. There is a mosaic of rice fields, usually in the lowest lands surrounding the river and stream valleys. Water is everywhere in the form of hill streams, flowing in different directions, and there are hundreds of little streams flowing down from the many hill ranges.

This is a major explanation as to how the animal exists in this region. I do think that the Mande-Burung uses these streams as routes, not just from A to B, but to feeding places. They are the easiest routes to follow in the thick jungle and are usually open, or not so overgrown. A tall person can walk for hundreds of miles upright without having to stoop. The streambeds are solid, in the way of rocks, sand and silt. The streams are always moving, and sluggish waterways are not as common, except the main rivers and their tributaries, of which there are many in the wider valleys. The rocks in the streams are mostly large. Gravel-sized stones are the base of many of the streams, with larger fist-sized rocks, then larger football-sized ones. Within these streams will be even larger chair- to car-sized rocks and some streams are just made up of the last two. The streambed is thus solid and easy to walk on, even if one is a very heavy animal such as a large human-like ape. Even buffalo can traverse many of the streams, yet many would be impassable for quadrupeds, but for bipedal animals, it would not be such a problem - especially if it can hold on with hands in slippery areas. And some areas are certainly slippery on the rocks, but with a tough bare foot, it would be easy. The streams are also cool, clear and full of life. Several species of freshwater crab and shrimps can be found - the crabs can be quite large and are easy to find sheltering under the rocks. Small fish are plentiful and so are amphibians. In some areas where there is still water, the populations of bullfrogs are immense. If one walks along the banks of even a small pond, typically over one hundred specimens may lunge into the water from the bank sides. The footprints of civets and small cats such as fishing cats are to be found along the silt banks. Many streams wind their way through bamboo forests, whilst others meander through plantations of bananas, tea, chilli, and many other kinds of farmed vegetables, but the majority of the hill streams run through virgin jungle. The streams are, quite simply, highways for certain animals.

Most of the reports of the Mande-Burung come from the streams themselves, or from areas not far from them or other rivers. It would make sense, therefore, to assume that these streams are central to the movements of the animal and its feeding habits.

It is debatable whether or not the Mande-Burung eats meat in the way of reptiles, birds or mammals. Would the animal attain all of its daily requirements from just vegetable matter? I think not. The animal is very bulky and tall. If such an animal were herbivorous, then it would have to spend a lot of time doing nothing but resting and sleeping. Would it actually need more protein in the form of animal life? I believe so. The nearest animals to the Mande-Burung that are solely herbivorous are the orangutans and gorillas, although both species are very different from each other in behaviour, physique and eating habits. Both species are not very mobile. The orangutan spends much of its time in the treetops doing little else but feeding, as the vegetation is basically low in protein. The gorilla is the same - it too spends most of the time digesting cellulose in the form of the vegetation it consumes. Although the gorilla is a social animal and lives in family groups, it does not spend a lot of time and energy moving any great distance within a day. This behaviour is typical of the large herbivores.

Chimpanzees on the other hand are omnivorous, and may have the luxury of meat several times a week. This meat is usually in the form of other primates or antelopes, and on a regular basis invertebrates in the form of ants and termites, and chimpanzees, on the other hand, do not loiter around for too long and are regularly on the move. If a very large primate needs to move around every day, often covering large distances, then it would make sense to eat a more protein full diet in the form of red meat. In addition, if the primate lives at high altitudes then temperatures would be a problem for a herbivore, so red meat would allow a healthy fat build up in wintertime. To carry a huge bulk of muscle around while travelling many miles in a bipedal way would almost certainly mean that the animals would need to eat as much meat as possible. This could mean a whole animal per week, perhaps in the form of a deer, or several marmots or other rodents - even birds and their eggs. It is believed that the Bigfoot of North America, as well as the large Yeti of Central Asia, often take cattle or yaks. There have been eyewitness accounts of them taking large animals and eating the flesh, or carrying the remains around with them, or stashing them in places.

While in Meghalaya, we did not hear of any accounts of the Mande-Burung taking cattle or goats, but I would not be surprised if accounts have been incorporated into literature or folk tales in the past. I do believe that the animal is omnivorous like us, and would not miss the opportunity to eat fresh meat in the form of mammals or birds. Smaller animals in the form of reptiles and fish would give the animal much needed protein, and if there were enough of these animals within its territory - if it indeed has one - then I am sure that this diet, along with much plant and fruits would sustain it. While on our search for the animal, we found remains of freshwater crabs that had been caught by the overturning of stream rocks. The crabs are rather small, but many of them, along with bamboo and insects could give all the daily requirements one animal needs. The Mande-Burung would not need to eat large bones, but small ones from the reptiles or small mammals would give it all its calcium requirements. If the villages found the remains of a goat eaten by a Mande-Burung for example, they might attribute the killing to a leopard or tiger, or one of the several other species of cats that could be living in the region. So far there have been no reports of the Mande-Burung attacking or eating humans, as there have been with a few of the other big man -ike apes in other parts of the world. If not, then maybe it has good reason to be shy of man, or perhaps it is not such a meat eater after all.

Maybe it does not travel large distances at all.

Perhaps I have been likening it to the Yeti of the higher, more remote areas of the Himalayas. After all, it would seem to be very similar in appearance. However, perhaps the fact that it lives in a more hospitable climate - with more in the way of fruit and small animals - it has less of a need for red meat.

The Garo tribes have always had a tradition of hunting, and many of the evergreen forest animals that existed have long since been hunted out of existence. This would have had an effect on the hunting ability of the Mande-Burung, and maybe it would have to eat more vegetation or fruit than usual, or maybe this has led it to eat more crustaceans and insects. Maybe in the past the animal only ate fruit, leaves and deer. It could well have been a scavenger and robbed the kills of tigers or leopards. Being supposedly a tall, massive and strong beast, in my opinion a Mande-Burung could probably come out best in a tussle with a tiger.

If we can imagine the area before modern humans arrived, then we can see a forest of great richness of biodiversity, and the Mande-Burung would have been the feared animal at the top of the food chain. Perhaps the species of deer and other primates did not fear it as much as the cunning leopard or the tiger, but I am sure that the animal would not have been able to sit within a herd of deer and been oblivious to them, like a scene of lowland gorillas in Africa. It is also strange that such a large animal would naturally be loners. I do think that this is due to the rarity of the beast and not by evolutionary choice. I can imagine them living in small family groups of perhaps two or three females with young, plus a male. They would share any meat and larger vegetation brought back by a male or female that may have gone out hunting. Maybe the animals did once stay for short periods and make a rough camp in one area that was rich in food before moving on. It is said that the animal makes a nest, and from one witness report, it would seem to be a neatly constructed shelter. The witness stated that, when he peered through the jungle at the sound of whimpering noises, he saw a distinct lean-to kind of shelter of intertwined bamboo stems and a small inner canopy within. In addition, the floor was cleared of debris and worn smooth. This would suggest that the animal has weaving skills more akin to early hominids.

The description of the animal would certainly seem to be of a slimmer animal than a gorilla and possess not so much of the body mass of many reports of the North American Bigfoot and the Asian Yeti. In my opinion it seems to resemble the Almasty of the Russian region more. After looking at the sketches made by eyewitnesses, it seems to have the slight look of orangutan about it. I wonder, then, if it might be related to Orang Pendek. The head seems to be domed in a similar way to that of gorillas, and this feature seems to be prevalent in most of the man-apes. The orangutan, on the other hand, does not have a domed head. The colour of the hair said to cover the whole of the body is brownish black, and the creature is said to have a distinctive musky smell. This also is common with the other alleged man-apes. According to reports, the hair is long, especially on the back of the head and neck. When pictures, or artists' impressions, of various apes and ape-men were shown to the eyewitnesses, it was interesting to watch each person carefully look at each picture and dismiss them one by one, except for the artists' impressions of Gigantopithecus. They would state that it looked like that, but with longer hair on its back of the head, neck and back. This is interesting, as this feature is not so apparent with the North American Bigfoot descriptions, and it certainly is not to be found in modern day chimpanzees or gorillas. However, it does have slight a similarity with orangutans, and I would assume certainly with Orang Pendek. This also has a similarity with humans, especially ancient hominids, as they too are more likely to have had long hair at the back of the head and neck.

The size of the footprints do not necessarily correspond to the height of the animal, but if it is slimmer than some of the other alleged man-apes, then it would be reasonable to assume that it is tall rather than broader. A giant foot does not mean that it has to be huge. We are simply comparing it to the modern human foot, in relation to body size. Most reports of the animal do seem to suggest that it stands upright at between six and ten feet tall. These judgements may be unreasonable considering that most witnesses are Garo people, and they may not have a very good sense of measurement as does Western man. Sizes of animals are usually exaggerated, as it looks good to impress, and this has happened in most tales or observations of cryptids and monsters. I am sure that the animal is tall, perhaps seven or eight feet. It may have a large foot in relation to body size, but, as our findings suggest - via footprints that we found - it would seem that the foot is broad, which would mean a broad leg, and – therefore - a broad body, similar to the depictions of man-apes in other regions of the world. It may seem to show that the animal does not walk while putting most pressure on to its heel. The footprints were deeper in the ground at the front of the foot region than the back. This also relates to findings of Bigfoot and other man-apes. It would seem that it can walk on its toes at times, despite not having a foot arch, or at least half walk on its toes. This may give it balance when walking on river bed silt, or it may give it grip on slippery rocks and mud, or it may give it speed and acceleration. The footprints also suggest that the animal has smaller toes than that of many of the alleged Bigfoot footprints, and even some of the Yeti footprints. There is certainly no opposable thumb as in the great apes. The animal obviously has not been a tree climber for a very long evolutionary time span. Reports also suggest a long figure and toe nails, maybe similar to humans but much thicker and stronger.

Most reports of the Mande-Burung are from the national parks of Nokrek and Balpakram as well as from Tura Peak. Tura Peak is relatively near to the town of Tura, in fact the peak dominates the town with its steep hillsides. The national parks are more remote and cover large areas, with little hunting in them. We did not find any evidence in Balpakram, but we did not go far and stayed on the well-worn track by vehicle first, then on foot for a small part of the way. There was evidence of elephant, buffalo and hog deer, also the scats of cats such as golden or clouded leopard. There were sandy areas of trackside alongside the very rocky terrain but no footprints of Mande-Burung. The land was dry and we experienced no rain, as - being November - it was still the dry season, and was technically autumn. During the monsoon season in June and July I would expect more footprints to be found, that is if the animals remain in the region and are not seasonably migrating. It would seem that most encounters are in the autumn and winter as opposed to the warmer and wetter summer months. This does not signify much in itself, but it may mean that the Garo people do not hunt so much in wet weather. Perhaps it is very dangerous to travel along the many small rivers and streams at such a time, and that may be why there are less reports. Most plants - bananas especially - fruit in the autumn time. However, this is a subtropical area so seasons in respect of fruiting plants are not that well defined, but there is a general rule. Global warming has also had its effects on the plant life and the winters are now much warmer than in previous years.

Meghalaya is very like an island in respect to its geography. It is a vast area that would fit snugly within southern England, yet the vast plains of Bangladesh to the south and west, and a large divide of rivers and mountains to the east - which really could be the most likely entry and exit into the region - surround it. To the north lies the great Brahmaputra River, which is very wide and acts as a natural barrier between Assam and Bhutan. It seem unlikely the Mande-Burung crosses these regions on a back and forth basis. If it is migrating solely on seasonal grounds such as to escape cooler temperatures from the north, then the animals would have to be coming through the north-east on a very long run through the northern highlands of Myanmar, on through the Nagaland before entering the East Kasi Hills district and then into the Garo Hills. It would seem a long haul for a large primate to do.

It is even more unlikely that the Mande-Burung comes down from Bhutan, as it would have to cross the mighty Brahmaputra River. This river is huge and very wide. It is more like an inland sea at most points, with only a few narrow areas in the dry season, yet it is shallow. It would not be impossible for the Mande-Burung to know the good crossing points of the several wide parts of the main river in the dry season and cross during the night at its lowest point. There are safe areas in which it could do so, but then it would be more exposed in the areas of Assam in-between. There may be corridors of forest connecting remote areas that could act as cover for them. There are also many bridges that the animals could cross, but as yet there are no reports of sightings. It may be that the animals remain in the Meghalaya area without ever needing to move away, but if they do then I would bet that they come from the north east. If indeed they only live in the Garo Hills then they would be severely inbred, as the number of the animals would seem to be very small. The Mande-Burung would then seem to be a relic, marooned on an island and doomed to extinction. I do not think that the Mande-Burung is quite the same as the Yeti, but a similar and smaller animal that could be, and maybe still is, found in Burma and Malaysia.

Whatever it is, it needs to be protected and fully studied. After meeting with eyewitnesses and seeing possible signs of the animal in the jungle, I can honestly say that I believe that the animal not only existed, but still exists albeit in very small numbers within Meghalaya.

We found possible footprints in at least two locations. The first set was in an area along a streambed where sightings had occurred earlier. We went down into a valley with one of the witnesses, and very soon the first print showed up, and then we found another two at least. All were facing the same direction, upstream. There was little doubt that what we were finding were footprints. We did not jump to any conclusions, but carefully debated all the possibilities. We firstly looked at the shape - a deep depression in the soft silt along the edge of the stream. It was pointing as if the maker had just walked from some large rocks and stepped towards the bank before turning to walk along the edge of the water, or to do a small detour alongside a lush area of vegetation that stretched for a few hundred yards out from the main stream bank. For approximately fifty yards the area upstream was overgrown so the animal may have done a short detour in exactly the same way as we did, for safety purposes. Whilst we judged the depression in the silt, it was clear that there were five toe-like depressions on the end of a foot shaped area. We looked firstly for a rock that could have been removed, but we found none and then wondered why that would happen anyhow. The impression was away from the main water and so if a human had moved a rock, it certainly was not to look for crabs. As there was no rock that was moved, how could a depression just appear in silt shaped like a giant foot?

It was about two inches deep into the silt, and when we stood in an area nearby and felt the mud, our weight dropped us down to only half the depth. This would imply that the animal was approximately twice our weight. The footprint was huge, but the same size as any other print made by a Yeti, but the toes were certainly very human-like. I am sure that it was not a hoax, as a hoaxer would have made several and made them more clear and obvious. This was not the case here. I have analysed many footprints of wild animals and I am used to looking at all the various forms in which an animal leaves its prints. The way in which the toes were lightly capped off the silty sand were in direct comparison to other animals in a similar situation. The area in which we were did not have many villages nearby, and there was just a few families living at the top of the gorge. We got the chance to speak to these families, and they did not show any signs of even knowing what we were doing, let alone being in a position to hoax Mande-Burung footprints in a haphazard, non-obvious manner.

We knew what we were looking at. It was as obvious as daylight. We were looking at the footprints of an unknown great ape, or hominid. We were all spellbound by our findings alongside the stream that day. The witness was surprised but did not seem to show any signs of sheepish grins or stardom in his eyes. He had seen the animal, and simply took us to where he had seen it, and that is when we found field evidence of it. When we spoke to the several witnesses, they were plainly being honest. They are not like Westerners, and they are not accustomed to hoaxing. They have no reason to. They are friendly tribes people who have just a bamboo hut and animals. Two of the witnesses felt so pleased that they had told us, as if a great burden had been removed from their shoulders; some may have been ridiculed by their peers as many villagers have not believed them, because with any legendary animal comes scepticism and awe. Many Garo folk are deeply superstitious, and see it as bad luck even to talk about the animal. Most villagers or townsfolk will believe that the animal did once exist, even if it does not now.

Like many of the larger mammal species that existed with them but have since disappeared from over-hunting, perhaps the Garo people see the Mande-Burung as being one of the same, but an animal that was not actually hunted by them. Despite this, the story of the severed hand in the market stall is intriguing.

A MISCELLANY OF INDIAN FORTEANA
OLL LEWIS

One day we hope to buy the CFZ a satellite phone so that we will be able to keep in daily touch with the expeditions, and update their progress on the CFZ blog. However, until then we are reliant on the problems of intermittent communication, and so—to keep the readers of the expedition blog entertained in the absence of any solid news—Oll Lewis penned a series of articles...

The Garo Hills

The India expedition should prove to be quite a contrast to conditions faced on many previous Centre for Fortean Zoology expeditions, not least because of the weather conditions I would expect the team to encounter in the Garo Hills. If you've followed our expeditions for a while you'll know that so far the CFZ has encountered almost anything nature can throw at them, from blistering heat with no precipitation to ice-capped mountains. One weather extreme not yet encountered, however, is constant rain.

I suspect this may be rectified soon as Adam, Richard, Chris, Dave and Jonathan make their way to the Garo Hills in their search for the Yeti, because the Garo Hills is one of the wettest places in the world. Some parts of the mountain range receive over 11 metres of rainfall a year, so I hope for their sakes they packed their waterproofs.

The Garo Hills are situated within the 41,700 square kilometres Meghalaya subtropical forests ecoregion in eastern India, which borders on the state of Assam, famous to many in the west because of its tea plantations. The Garo Hills region is a subtropical moist broadleaf forest eco-region home to around 320 different orchids and the beautiful *Nepenthes khasiana* pitcher plant.

The plant - whose local name among the A·chik Mande people of the Garo Hills is 'memang-koksi', which means 'basket of the devil' - is endangered but shows great genetic diversity and there are several cultivation projects in progress to help ensure the rare plants survival. Animals in the area include Asian elephant *(Elephas maximus)*, Asiatic wild dog *(Cuon alpinus)*, sun bear *(Ursus malayanus)*, sloth bear *(Melursus ursinus)*, smooth-coated otter *(Lutrogale perspicillata)*, Indian civet *(Viverra zibetha)*, Chinese pangolin *(Manis pentadactyla)*, Indian pangolin *(Manis crassicaudata)*, Assamese macaque *(Macaca assamensis)*, bear macaque *(Macaca arctoides)*, capped leaf monkey *(Semnopithecus pileatus)*, and Hoolock gibbon *(Hylobates hoolock)*, tiger *(Panthera tigris)* and clouded leopard *(Pardofelis nebulosa)*. Some of these, like the sun bear for example, have been offered as non-cryptid explanations for Yeti sightings in the area.

Kallana, The Cryptid Dwarf Elephant

People of the nomadic Kani tribes of India's Western Gnats have long talked of an animal they call Kallana. According to the indigenous people there are two morphologically distinct groups of elephants in the Peppara forest range, the first of which is the common Indian elephant. But the other is a dwarf variety they call the Kallana. There has been much debate as to whether the Kallana exist at all and expeditions to the area to search for the creatures have come back empty-handed, as many expeditions that search for cryptids do. The thing is, despite what some people would have you believe, finding an animal known only from anecdotal evidence is not as easy as simply pushing a pin in a map and saying "We'll search here, come on lads! To the Kongamato-mobile!" Hidden animals tend to be hidden for a reason and to discover such a creature if it exists one has to have the pure dumb luck of being in the right place at the right time, no matter how good a tracker you are.

The Kani people were, despite increasing scepticism from some nay-sayers, quite adamant that the Kallana exist because several of them had seen and observed the diminutive elephants with their own eyes. According to the Kani the Kallana grow to no more than about 1.5 metres in height (5ft), have long hairy tails that reach to the ground and avoid Indian elephant groups never mixing with them and often going out of their way to avoid encountering them. The peculiar pygmy pachyderms are said to live on a diet of grass, bamboo leaves, tubers and bark and be able to climb steep rocky inclines that larger elephants would have great difficulty with.

It remains to be seen whether the Kallana are a wholly different species of elephant from the Indian elephant. They are most likely just a different variety that has been fairly genetically isolated. But one thing looks likely... they do exist. A photograph was taken of one of the creatures in, according to conflicting reports, either 2005 or 2010 in the Peppara Wildlife Sanctuary.

The Ancient Sea-monsters of India

Monsters and strange animals have, like in many places around the world, been reported in and around India for thousands of years. A particular type of sea monster, known in the west as Ketea Indikoi (literally 'Indian sea-monsters'), was reported to inhabit the Indian Ocean near Sri Lanka. The monsters supposedly could be encountered in a variety of forms, usually with the head of one animal and the body of a fish. The animals' heads were often said to be mammalian ones ranging from lions to rams, but as well as these there were claims that some even had the heads of women with spines instead of hair. The creatures are also said to have been able to live on land as well as in the oceans.

The animals, if they existed at all, were likely just the result of travellers distorting descriptions of perfectly normal animals found in the area, like large crocodiles. The most well known description of these creatures dates back to the second century AD in the part fable, part natural history work, *De Natura Animalium* (On the nature of animals) by Claudius Aelianus, which is based on reports that would have passed through many different people before reaching his ears.

> 'Those [Indians of Taprobane--\ modern day Sri Lanka] that live near to the sea . . . devote themselves to catching fish and sea-monsters (ketoi). For they assert that the sea which surrounds the circuit of their island breeds a multitude past numbering of fishes and monsters, and moreover that they have the heads of lions and leopards and wolves and rams, and, still more wonderful to relate, that there are some which have the forms of Satyroi with the faces of women, and these have spines attached in place of hair. They tell of others too which have strange forms whose appearance not even men skilled in painting and in combining bodies of diverse shapes to make one marvel at the sight, could portray with accuracy or represent for all of their artistic skill; for these creatures have immense and coiling tails, while for feet they have claws or fins. I learn too that they are amphibious and that at night they graze the fields, for they eat the grass as cattle and rooks do; they enjoy the ripe fruit of the date-palm and therefore shake the trees with their coils, which being supple and capable of embracing, they fling round them. So when the shower of dates has fallen because of this violent shaking, they feed upon it. And then as the night wanes and before it is clear daylight these creatures plunge into the ocean and disappear as the dawn begins to glow.'

Make of that what you will....

The Little Known Garo Hills Caecilian

As well as the various mammal species I've mentioned in previous blogs, the Garo Hills are, as one would expect from such a wet part of the world, home to several unique species of amphibian. There is little known about many of the amphibians endemic to the area, some of which are only known to exist from just one or two specimens. This does not mean that the species is particularly rare but that it is probably either elusive or just won't be encountered unless you search in a specific place. For example, we all know that earthworms exist and are very common, but unless you dig into the soil or go outside after heavy rain on a regular basis then you seldom encounter them.

It is, however, difficult to judge the reasons why an animal is seldom encountered, which could be due to a number of factors including low numbers, a fragmented population or by simply looking in the wrong places, and in these cases, especially when habitat is threatened, the creature is listed as data deficient in the IUCN's Red List in the hope that more will found out about the creature in future and the entry can be updated with a more appropriate category.

One such species is *Ichthyophis garoensis*, the Garo Hills caecilian. The caecilian was described by Pillai and Ravichandran in 1999 and is known from only two specimens. The caecilian is seldom encountered as it lives in moist leaf litter, apparently venturing to the surface very rarely indeed. According to the Red List (http://www.iucnredlist.org/apps/redlist/details/59617/0) the species is known to inhabit the Anogiri Lake area of the Garo Hills and may also be found in Assam, where several specimens thought to be of the same species have been spotted. Because the only known specimens were both been found close to water one assumption is that the caecilians are aquatic in their larval stage. Like many species that live in the Garo Hills the species is potentially under threat from loss of habitat through logging or forest clearances, so unless humanity is careful we could lose this enigmatic species before we learn much more about it.

The Wolf Children of Calcutta

Something that, thanks to Kipling's character Mowgli, will forever be associated with India are stories of feral children. The story of Mowgli, told through the short story 'in the Rukh', where the character is first introduced as an adult, and the Jungle Books, which tell of Mowgli's childhood in the Indian jungle, deal with the story of how Mowgli, after surviving a tiger attack at a young age, was raised and protected by animals; notably wolves, a formerly domesticated panther and a bear. Mowgli's story was partially based on the stories and legends of feral children raised by animals found throughout the world in many different cultures, and in Europe the concept of children being raised by wolves dates as far back as the foundation myths of the city of Rome by Romulus and Remus in 753 BC.

In India often a child having been raised by wolves was used as a standby explanation for why a child had behaviour diverging from what would be considered normal; for example many who may have been on the autistic spectrum were unfairly characterised as feral children in some parts of the country. One famous case was that of Amala and Kamala, two children, allegedly raised by wolves, who found their way to the orphanage of Reverend Joseph Amrito Lal Singh in Calcutta.

According to the first stories Rev. Singh told, the girls were given to him by a man in Godamuri, but this story was later changed. In the revised version of the story Singh claimed to have rescued the two young girls from the wolves' den itself in 1920 when Amala was about 18 months and Kamala about 8 years old. Singh named the girls and wrote about their progress in a diary.

Singh recorded the girls' lives at his orphanage in meticulous detail, telling of how both girls seemed to possess traits inherited from their lupine upbringing, such as never wanting to dress, a nocturnal lifestyle and howling at night in a vain attempt to call to their pack. The girls also had hard skin on their knees and hands from walking on all fours, would not stand upright and had excellent night vision. Interestingly, though they refused cooked meat and would scratch and bite at anyone attempting to feed them, the girls would eat raw meat from bowls. The fact that they ate from bowls like a pet dog seems somewhat at odds with the rest of the wild wolf-like traits they were said to have.

EXPEDITION REPORT: India 2010

Amala died a few months after entering the orphanage from a kidney infection and Kamala lived to be about 17 years old, dying in 1929, also from a kidney infection, by which time she had been taught to walk upright and had a limited human vocabulary.

There are several aspects of the tale that cast doubt on the entire story and these were identified by Serge Aroles in the book *L'Enigme des enfants-loup*. The diaries were actually written in 1936, photographs of the wolf children were taken in 1937 (8 years after Kamala's death) with local children posing as the girls, and according to doctors from the orphanage, Kamala did not walk around on all fours or have any hard skin from having done so.

More disturbingly, several eyewitnesses claimed that Singh used to beat Kamala in order to get her to perform for visitors. This puts a whole new spin on the case and it could well be that rather than Kamala's behaviour being due to being raised by wolves, it could have been the result of neglect and abuse in her early years like the American "feral child" known as Genie who was discovered in 1970.

Genie displayed similar animal-like traits and was raised not by wolves but alone in an enclosed crib and tied to a chair during waking hours up until she was discovered when she was about 13. Such social isolation resulted in, among other things, an extremely limited vocabulary and unusual way of walking. It has been alleged that Singh fabricated or exaggerated the girls' feral nature in order to acquire more funds for his orphanage, which was in a dire financial situation at the time.

Living On Air

Not everything unusual that happens in India is cryptozoological in nature; there are a great many Fortean feats claimed by Indian people as well. One man reputed to be able to perform seemingly impossible tasks is Prahlad Jani, a man who claims to live on air alone.

Jani was born Chunriwala Mataji in the Indian state of Gujarat in 1929 and left home at the age of seven to live in the jungle. It was while he was in the jungle when he was eleven years old that he claims to have had a religious experience after which he became a devoted follower of the Hindu goddess Amba. Afterwards he dressed as a female with long hair to show his devotion for the goddess. Jani claims that the goddess blesses him with an elixir called amrit, which flows out of a hole in his pallet, and because of this divine elixir's powers he does not need to consume food or water, and needs air alone to survive. He has been living as a hermit in a rainforest cave since the 1970s.

Recognising the potential advantages of being able to live without food or water for soldiers and astronauts if the claims were in any way true, Professor Sundhir Shah of the Sheth K.M. School of Post Graduate Medicine and Research, Sheth V. S. General Hospital, Ahmedabad, decided to study Prahlad Jani in both 2003 and 2010.

In the 2003 tests in Sterling Hospital, Ahmedabad, it was established that Jani was physically normal apart from a hole in his pallet (this being the hole through which he claim's the elixir flows), and he was kept under observation. Jani was not observed drinking and the only water he was given by the scientists was 100ml of water to use as a mouthwash each day. Scientists observed urine forming in his bladder but he was not observed passing urine or stools during the ten days he was kept under observation at the hospital.

Although Jani did not engage in strenuous exercise during the ten-day trial he did lose weight, which cast doubt on his claims to be able to survive indefinitely on air alone. During the 2010 trial at the same hospital Jani was kept under observation by CCTV and given regular blood tests, again scientists observed fluctuating amounts of water in his bladder but did not see him pass water. Sanal Edamaruku; a sort of Indian version of Richard Dawkins but with a background in political study rather than science; described the experiment as a farce because it had been possible for Jani to move outside of the CCTV's field of view and because he, Edamaruku, had not been given permission to inspect the project during its operation. Quite why a man with no scientific training would expect given access to an expensive experiment when he clearly intended to rubbish it whether the project deserved being rubbished or not, I have no idea.

Jani may well be able to go without food or water for long periods of time as he claims or he may be pulling off a bit of David-Blaine-style trickery. Without further periods of testing it is hard to say for sure but if it is a genuine talent then all humans may be capable of similar feats with the right amount of training or practice. Personally, I am not particularly convinced that anyone could live without food and water indefinitely but you do hear some remarkable stories of survival where a person has performed similar feats, so a few days could be well within the realms of possibility.

The New Delhi Monkey Man Panic

In May 2001 residents of New Delhi came under attack from a strange creature. The creature, if indeed it was a creature, caused more harm to the population due to the mass panic that followed in its wake than it was personally responsible for. The animal was dubbed 'The Monkey Man' by the media and differing reports as to the monkey man's appearance surfaced in the city.

The police described it as 4-foot-6 with a dark coat of hair, but some eyewitnesses described a

EXPEDITION REPORT: India 2010

The police say the creature is 4'6", wears only a dark coat of hair

Eyewitness says it is 5'6", wears black and sports a helmet, with shining red eyes

very different-sounding beast describing it as 5-foot-6 and wearing black clothes with a sports helmet and glowing red eyes. If the eyewitness statements were truthful then this would suggest that rather than a monkey, the creature was a man hoping to cause panic by dressing in a sinister manner and harassing people, similar to Spring-heeled Jack in Victorian London.

The first time the monkey man was seen was on the 13th of May 2001 when he was allegedly responsible for minor injuries to 15 people, including bites scratches and bruises. This was the only day upon which the monkey man is said to have actually physically harmed anyone; all other accidents, injuries and even the following deaths were caused by the mass panic that followed in the wake of these incidents.

Within two days fear had taken hold of the local populous and a pregnant woman fell to her death in a stairwell after neighbours saw her running and shouting about having seen the monkey man. The next casualty would be a four-foot-tall wandering Hindu mystic who was beaten by a crowd who mistook him for the monkey man, and a similar fate befell a van driver who sustained multiple fractures during his beating.

After a few days without incident the monkey man and the resulting panic he had caused was quietly forgotten and the police did not arrest anyone for the panic, subscribing to the theory that the initial attacks that caused the panic were likely a combination of animal attacks and in some cases just common accidents. Further sightings of the monkey man were made in Kanpur, Uttar Pradesh, in February 2002 and in New Delhi once more in July 2002, but as it has not been seen since it is probable that whatever caused the panic of 2001, creature or man, it has probably moved on.

Take me to the River

One afternoon in February 1973 a priest out for an afternoon walk by the Kuano river near the village of Baragdava, Uttar Pradesh in northern India, saw a most peculiar sight. The priest was crossing the river via a dam when he saw a naked boy of about 15 years old walking into the water and diving under the waters for about a minute before rising up triumphantly with a fish in his mouth. The lad ate the fish raw then swam away. When the priest told the villagers of this unusual sight a woman named Somni told him how her son Ramchandra was swept away by the river when he was a baby and would be roughly the same age as the boy.

Although another villager saw the boy by the river a few days later the boy was not seen again in the area until 1978 when Somni spotted him in a field. Somni was able to look at a birthmark on the boy and confirmed it was her missing son before he ran off. This time the villagers were more prepared and managed to find and capture the boy to take him back to Baragdava. He eventually escaped again and went back to living in the river but having learned that humans would feed him and meant him no harm, he was no longer afraid of human contact.

Ramchandra would occasionally come to the village or regularly be spotted by the river and seemed quite happy with his lot in life. It was on one of his visits into the village that the boy was observed by a journalist who wrote up the story of the feral child (who was by now a man) for the Allahabad magazine and observed Ramchandra submerging himself underwater for much longer than a normal person would be capable of. The journalist said that Ramchandra had dark skin with a subtle green tinge to it and walked upright but with a quite clumsy gait. He also had very hard skin on his feet as one might expect from someone who spent their life outdoors and without the comfort of shoes. It is thought that the boy was deaf and certainly was never observed speaking or making any human sounds.

Compared to most feral children it would seem that Ramchandra led quite a nice life and it doesn't look like he was ever persecuted, abused or exploited in order to make money or make somebody famous, as is the fate of most feral children. His death, however, was tragic. He had made his way into the nearby village of Sanrigar where a woman threw a large amount of boiling water at him in fright. The woman, unaware that this was the river boy from the next village along, had possibly been frightened by Ramchandra's appearance, strange walk and the fact that he would have been completely naked; and she may have feared that she was about to be sexually attacked by a madman. Ramchandra ran back to the river that he knew and loved but it was too late for the young fish lad and his blistered body was found in the river later that day.

The Disappearing Western Hoolock

As you will have seen from some of the previous articles I've written for this blog the Garo Hills is home to a large number of rare species. One of those species, and indeed a species that some people have postulated may be responsible for some sightings of the Indian Yeti, is the western hoolock gibbon *(Hoolock hoolock)*. The western hoolock is an endangered species of tail-less gibbon and can grow to a maximum size of around 90cm, making it, in terms of size of individuals, the second largest known species of gibbon in the world.

The western hoolock is under considerable pressure from several different sources, including deforestation and hunting, and as a result the population has declined by 50% in the last 40 years. Unless conservation efforts pay off this trend is likely to continue. As the population becomes more fragmented the speed of decline will only accelerate as groups of the gibbons become more and more isolated from each other. This isolation of groups will lead to smaller local gene-pools and the inevitable loss of potential genetic combinations that could reduce resistance to disease and other environmental pressures, and if this isolation is caused by deforestation and the fragmentation of habitat then, if one local group is wiped out it is less likely that the area will be recolonised naturally. There are estimated to be less than 5000 individual western hoolock gibbons across their entire range in Bangladesh, north eastern India and north western Myanmar. It is also thought to occur in parts of the Tibet area of China, and it has been listed among the world's 25 most at risk primates by the IUCN.

Another potential threat to the western hoolock's survival is its close relative the eastern hoolock gibbon *(Hoolock leuconedys)*, which is also a threatened species but classed as vulnerable rather than the more threatened endangered western hoolock. Typically the eastern hoolock is only found to the east of the Chindwin River but the two species are not reproductively isolated and hybridisation between the two species occurs towards the river's source. It is known that there is a population of the eastern hoolock, within the range of the western hoolock, in Arunachal Pradesh and it is certainly a possibility that hybridisation will have occurred around the edges of this population as well. Isolated hybridisation is one thing but if the eastern hoolock encroaches further into the western hoolock's territory there is a danger that hybridisation could spread, which would further impact the population of western hoolocks.

It is certainly not too late to save the western hoolock and along with a CITES listing there are several habitat conservation projects in India from which the western hoolock benefits, but in order to halt the decline of the species and even reverse that trend a lot more will have to be done.

Here Comes The Sun Bear

One of the creatures often used to explain away sightings of the Indian yeti is the sun bear *(Helarctos malayanus)*. It is easy to see how in some cases, for example when a the 'yeti' is glimpsed very briefly in pitch darkness or from behind, a mistaken identification like this

EXPEDITION REPORT: India 2010

could be possible, after all a lot of possible lake monster sightings can turn out to have been caused by a floating log or a perfectly natural wave. Some cases... but not all. When a sun bear is seen from the front there is no mistaking the distinctive orange horse-shoe marking under the chin and its creamy-coloured, distinctly bear-like face. I find it hard to believe that anybody who managed to get a half decent look at a sun bear would be insistent that they had seen a cryptid. This especially applies to locals who would certainly be familiar with most of the local wildlife, not least because a lot of it; like tigers, leopards, snakes and indeed sun bears; is potentially very dangerous indeed.

That is not to say that sun bears are not interesting animals in their own right. Sun bears are the smallest of the Ursidaes, measuring around 1.2 metres (4ft) in length. Sun bears will often climb trees to find safe places to rest during the day, and are nocturnal. They are classed as vulnerable by the IUCN due to deforestation fragmenting the species' habitat throughout their range and uncontrolled exploitation in trade of body parts for Chinese medicine amongst other things.

EPILOGUE: DISAPPOINTING NEWS

Posted by Jon Downes on the CFZ Blog, Thursday, June 16, 2011

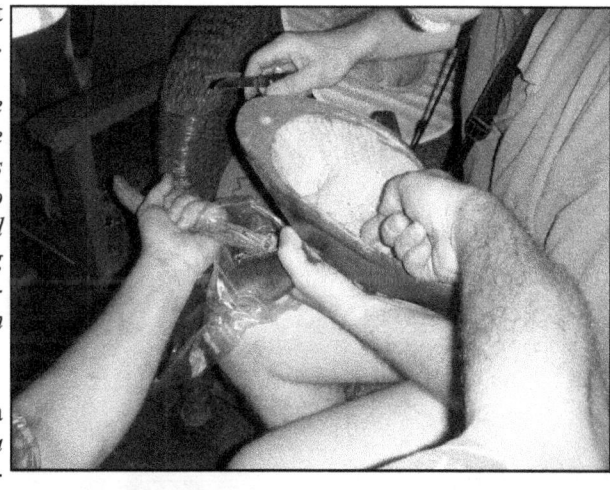

I received an e-mail late last night from Lars Thomas.

"Hi Guys, Finally we have the results of the DNA analysis of the antler samples from India. It has taken an awful lot of time, but we do need to check and recheck and check again - and earn a living every now and then :-). Unfortunately there was no new species in there after all... "

The antlers turned out to be from a Sambhur (Sambar) deer - (*Rusa unicolor*) presumably a juvenile - because although the antlers of an adult resemble those of a fallow deer, the antlers of the juvenile surprised me greatly by looking like those of a muntjac. It is mildly disappointing but our job is to find out the truth, and not purely to look for cryptids, and we have found the truth.

Richard Freeman and I would like to thank Tom Gilbert and his team for all their painstaking work. And as I often do, I am going to take refuge in the words of Rudyard Kipling:

As the dawn was breaking the Sambhur belled --
Once, twice and again!
And a doe leaped up, and a doe leaped up
From the pond in the wood where the wild deer sup.
This I, scouting alone, beheld,
Once, twice, and again!

STILL ON THE TRACK OF UNKNOWN ANIMALS

The Centre for Fortean Zoology, or CFZ, is a non profit-making organisation founded in 1992 with the aim of being a clearing house for information, and coordinating research into mystery animals around the world.

We also study out of place animals, rare and aberrant animal behaviour, and Zooform Phenomena; little-understood "things" that appear to be animals, but which are in fact nothing of the sort, and not even alive (at least in the way we understand the term).

Not only are we the biggest organisation of our type in the world, but - or so we like to think - we are the best. We are certainly the only truly global cryptozoological research organisation, and we carry out our investigations using a strictly scientific set of guidelines. We are expanding all the time and looking to recruit new members to help us in our research into mysterious animals and strange creatures across the globe.

Why should you join us? Because, if you are genuinely interested in trying to solve the last great mysteries of Mother Nature, there is nobody better than us with whom to do it.

Members get a four-issue subscription to our journal *Animals & Men*. Each issue contains nearly 100 pages packed with news, articles, letters, research papers, field reports, and even a gossip column! The magazine is Royal Octavo in format with a full colour cover. You also have access to one of the world's largest collections of resource material dealing with cryptozoology and allied disciplines, and people from the CFZ membership regularly take part in fieldwork and expeditions around the world.

The CFZ is managed by a three-man board of trustees, with a non-profit making trust registered with HM Government Stamp Office. The board of trustees is supported by a Permanent Directorate of full and part-time staff, and advised by a Consultancy Board of specialists - many of whom are world-renowned experts in their particular field. We have regional representatives across the UK, the USA, and many other parts of the world, and are affiliated with other organisations whose aims and protocols mirror our own.

You'll find that the people at the CFZ are friendly and approachable. We have a thriving forum on the website which is the hub of an ever-growing electronic community. You will soon find your feet. Many members of the CFZ Permanent Directorate started off as ordinary members, and now work full-time chasing monsters around the world.

Write to us, e-mail us, or telephone us. The list of future projects on the website is not exhaustive. If you have a good idea for an investigation, please tell us. We may well be able to help.

We are always looking for volunteers to join us. If you see a project that interests you, do not hesitate to get in touch with us. Under certain circumstances we can help provide funding for your trip. If you look on the future projects section of the website, you can see some of the projects that we have pencilled in for the next few years.

In 2003 and 2004 we sent three-man expeditions to Sumatra looking for Orang-Pendek - a semi-legendary bipedal ape. The same three went to Mongolia in 2005. All three members started off merely subscribers to the CFZ magazine. Next time it could be you!

We have no magic sources of income. All our funds come from donations, membership fees, and sales of our publications and merchandise. We are always looking for corporate sponsorship, and other sources of revenue. If you have any ideas for fund-raising please let us know. However, unlike other cryptozoological organisations in the past, we do not live in an intellectual ivory tower. We are not afraid to get our hands dirty, and furthermore we are not one of those organisations where the membership have to raise money so that a privileged few can go on expensive foreign trips. Our research teams, both in the UK and abroad, consist of a mixture of experienced and inexperienced personnel. We are truly a community, and work on the premise that the benefits of CFZ membership are open to all.

Reports of our investigations are published on our website as soon as they are available. Preliminary reports are posted within days of the project finishing.

Each year we publish a 200 page yearbook containing research papers and expedition reports too long to be printed in the journal. We freely circulate our information to anybody who asks for it.

We have a thriving YouTube channel, CFZtv, which has well over two hundred self-made documentaries, lecture appearances, and episodes of our monthly webTV show. We have a daily online magazine, which has over a million hits each year.

Each year since 2000 we have held our annual convention - the Weird Weekend. It is three days of lectures, workshops, and excursions. But most importantly it is a chance for members of the CFZ to meet each other, and to talk with the members of the permanent directorate in a relaxed and informal setting and preferably with a pint of beer in one hand. Since 2006 - the Weird Weekend has been bigger and better and held on the third weekend in August in the idyllic rural location of Woolsery in North Devon.

Since relocating to North Devon in 2005 we have become ever more closely involved with other community organisations, and we hope that this trend will continue. We have also worked closely with Police Forces across the UK as consultants for animal mutilation cases, and we intend to forge closer links with the coastguard and other community services. We want to work closely with those who regularly travel into the Bristol Channel, so that if the recent trend of exotic animal visitors to our coastal waters continues, we can be out there as soon as possible.

Apart from having been the only Fortean Zoological organisation in the world to have consistently published material on all aspects of the subject for over a decade, we have achieved the following concrete results:

• Disproved the myth relating to the headless so-called sea-serpent carcass of Durgan beach in Cornwall 1975
• Disproved the story of the 1988 puma skull of

Lustleigh Cleave
- Carried out the only in-depth research ever into the mythos of the Cornish Owlman.
- Made the first records of a tropical species of lamprey
- Made the first records of a luminous cave gnat larva in Thailand
- Discovered a possible new species of British mammal - the beech marten
- In 1994-6 carried out the first archival fortean zoological survey of Hong Kong
- In the year 2000, CFZ theories were confirmed when a new species of lizard was added to the British List
- Identified the monster of Martin Mere in Lancashire as a giant wels catfish
- Expanded the known range of Armitage's skink in the Gambia by 80%
- Obtained photographic evidence of the remains of Europe's largest known pike
- Carried out the first ever in-depth study of the ninki-nanka
- Carried out the first attempt to breed Puerto Rican cave snails in captivity
- Were the first European explorers to visit the `lost valley` in Sumatra
- Published the first ever evidence for a new tribe of pygmies in Guyana
- Published the first evidence for a new species of caiman in Guyana
- Filmed unknown creatures

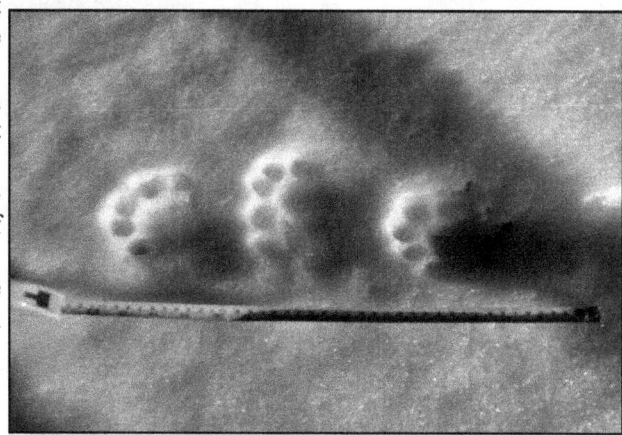

on a monster-haunted lake in Ireland for the first time
- Had a sighting of orang pendek in Sumatra in 2009
- Found leopard hair, subsequently identified by DNA analysis, from rural North Devon in 2010
- Brought back hairs which appear to be from an unknown primate in Sumatra
- Published some of the best evidence ever for the almasty in southern Russia

CFZ Expeditions and Investigations include:

- 1998 Puerto Rico, Florida, Mexico (Chupacabras)
- 1999 Nevada (Bigfoot)
- 2000 Thailand (Naga)
- 2002 Martin Mere (Giant catfish)
- 2002 Cleveland (Wallaby mutilation)

- 2003 Bolam Lake (BHM Reports)
- 2003 Sumatra (Orang Pendek)
- 2003 Texas (Bigfoot; giant snapping turtles)
- 2004 Sumatra (Orang Pendek; cigau, a sabre-toothed cat)
- 2004 Illinois (Black panthers; cicada swarm)
- 2004 Texas (Mystery blue dog)
- 2004 Loch Morar (Monster)
- 2004 Puerto Rico (Chupacabras; carnivorous cave snails)
- 2005 Belize (Affiliate expedition for hairy dwarfs)
- 2005 Loch Ness (Monster)
- 2005 Mongolia (Allghoi Khorkhoi aka Mongolian death worm)

- 2006 Gambia (Gambo - Gambian sea monster, Ninki Nanka and Armitage's skink
- 2006 Llangorse Lake (Giant pike, giant eels)
- 2006 Windermere (Giant eels)
- 2007 Coniston Water (Giant eels)
- 2007 Guyana (Giant anaconda, didi, water tiger)
- 2008 Russia (Almasty)
- 2009 Sumatra (Orang pendek)
- 2009 Republic of Ireland (Lake Monster)
- 2010 Texas (Blue Dogs)
- 2010 India (Mande Burung)

For details of current membership fees, current expeditions and investigations, and voluntary posts within the CFZ that need your help, please do not hesitate to contact us.

The Centre for Fortean Zoology,
Myrtle Cottage,
Woolfardisworthy,
Bideford, North Devon
EX39 5QR

Telephone 01237 431413
Fax +44 (0)7006-074-925
eMail info@cfz.org.uk

Websites:

www.cfz.org.uk
www.weirdweekend.org

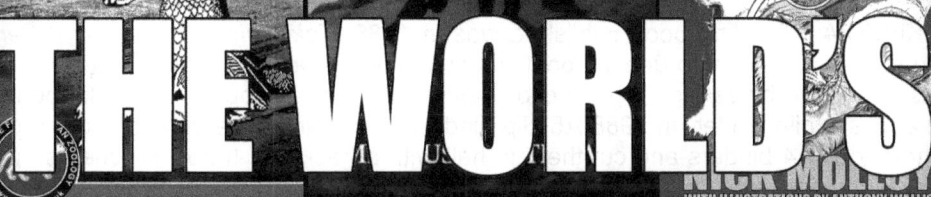

THE WORLD'S WEIRDEST PUBLISHING COMPANY

HOW TO START A PUBLISHING EMPIRE

Unlike most mainstream publishers, we have a non-commercial remit, and our mission statement claims that "we publish books because they deserve to be published, not because we think that we can make money out of them". Our motto is the Latin Tag *Pro bona causa facimus* (we do it for good reason), a slogan taken from a children's book *The Case of the Silver Egg* by the late Desmond Skirrow.

WIKIPEDIA: "The first book published was in 1988. *Take this Brother may it Serve you Well* was a guide to Beatles bootlegs by Jonathan Downes. It sold quite well, but was hampered by very poor production values, being photocopied, and held together by a plastic clip binder. In 1988 A5 clip binders were hard to get hold of, so the publishers took A4 binders and cut them in half with a hacksaw. It now reaches surprisingly high prices second hand.

The production quality improved slightly over the years, and after 1999 all the books produced were ringbound with laminated colour covers. In 2004, however, they signed an agreement with Lightning Source, and all books are now produced perfect bound, with full colour covers."

Until 2010 all our books, the majority of which are/were on the subject of mystery animals and allied disciplines, were published by `CFZ Press`, the publishing arm of the Centre for Fortean Zoology (CFZ), and we urged our readers and followers to draw a discreet veil over the books that we published that were completely off topic to the CFZ.

However, in 2010 we decided that enough was enough and launched a second imprint, `Fortean Words` which aims to cover a wide range of non animal-related esoteric subjects. Other imprints will be launched as and when we feel like it, however the basic ethos of the company remains the same: Our job is to publish books and magazines that we feel are worth publishing, whether or not they are going to sell. Money is, after all - as my dear old Mama once told me - a rather vulgar subject, and she would be rolling in her grave if she thought that her eldest son was somehow in `trade`.

Luckily, so far our tastes have turned out not to be that rarified after all, and we have sold far more books than anyone ever thought that we would, so there is a moral in there somewhere...

Jon Downes,
Woolsery, North Devon
July 2010

CFZ PRESS

Other Books in Print

Weird Waters—
The lake and sea monsters of Scandinavia and the Baltic States by Lars Thomas
The Inhumanoids by Barton Nunnelly
Monstrum! A Wizard's Tale by Tony "Doc" Shiels
CFZ Yearbook 2011 edited by Jonathan Downes
Karl Shuker's Alien Zoo by Shuker, Dr Karl P.N
Tetrapod Zoology Book One by Naish, Dr Darren
The Mystery Animals of Ireland by Gary Cunningham and Ronan Coghlan
Monsters of Texas by Gerhard, Ken
The Great Yokai Encyclopaedia by Freeman, Richard
NEW HORIZONS: Animals & Men *issues 16-20 Collected Editions Vol. 4*
by Downes, Jonathan
A Daintree Diary -
Tales from Travels to the Daintree Rainforest in tropical north Queensland, Australia
by Portman, Carl
Strangely Strange but Oddly Normal by Roberts, Andy
Centre for Fortean Zoology Yearbook 2010 by Downes, Jonathan
Predator Deathmatch by Molloy, Nick
Star Steeds and other Dreams by Shuker, Karl
CHINA: A Yellow Peril? by Muirhead, Richard
Mystery Animals of the British Isles: The Western Isles by Vaudrey, Glen
Giant Snakes - Unravelling the coils of mystery by Newton, Michael
Mystery Animals of the British Isles: Kent by Arnold, Neil
Centre for Fortean Zoology Yearbook 2009 by Downes, Jonathan
CFZ EXPEDITION REPORT: Russia 2008 by Richard Freeman *et al*, Shuker, Karl (fwd)
Dinosaurs and other Prehistoric Animals on Stamps - A Worldwide catalogue
by Shuker, Karl P. N
Dr Shuker's Casebook by Shuker, Karl P.N
The Island of Paradise - chupacabra UFO crash retrievals,
and accelerated evolution on the island of Puerto Rico by Downes, Jonathan
The Mystery Animals of the British Isles: Northumberland and Tyneside by Hallowell, Michael J
Centre for Fortean Zoology Yearbook 1997 by Downes, Jonathan (Ed)

Centre for Fortean Zoology Yearbook 2002 by Downes, Jonathan (Ed)
Centre for Fortean Zoology Yearbook 2000/1 by Downes, Jonathan (Ed)
Centre for Fortean Zoology Yearbook 1998 by Downes, Jonathan (Ed)
Centre for Fortean Zoology Yearbook 2003 by Downes, Jonathan (Ed)
In the wake of Bernard Heuvelmans by Woodley, Michael A
CFZ EXPEDITION REPORT: Guyana 2007 by Richard Freeman *et al*, Shuker, Karl (fwd)
Centre for Fortean Zoology Yearbook 1999 by Downes, Jonathan (Ed)
Big Cats in Britain Yearbook 2008 by Fraser, Mark (Ed)
Centre for Fortean Zoology Yearbook 1996 by Downes, Jonathan (Ed)
THE CALL OF THE WILD - Animals & Men issues 11-15
Collected Editions Vol. 3 by Downes, Jonathan (ed)
Ethna's Journal by Downes, C N
Centre for Fortean Zoology Yearbook 2008 by Downes, J (Ed)
DARK DORSET -Calendar Custome by Newland, Robert J
Extraordinary Animals Revisited by Shuker, Karl
MAN-MONKEY - In Search of the British Bigfoot by Redfern, Nick
Dark Dorset Tales of Mystery, Wonder and Terror by Newland, Robert J and Mark North
Big Cats Loose in Britain by Matthews, Marcus
MONSTER! - The A-Z of Zooform Phenomena by Arnold, Neil
The Centre for Fortean Zoology 2004 Yearbook by Downes, Jonathan (Ed)
The Centre for Fortean Zoology 2007 Yearbook by Downes, Jonathan (Ed)
CAT FLAPS! Northern Mystery Cats by Roberts, Andy
Big Cats in Britain Yearbook 2007 by Fraser, Mark (Ed)
BIG BIRD! - Modern sightings of Flying Monsters by Gerhard, Ken
THE NUMBER OF THE BEAST - Animals & Men issues 6-10
Collected Editions Vol. 1 by Downes, Jonathan (Ed)
IN THE BEGINNING - Animals & Men issues 1-5 Collected Editions Vol. 1 by Downes, Jonathan
STRENGTH THROUGH KOI - They saved Hitler's Koi and other stories by Downes, Jonathan
The Smaller Mystery Carnivores of the Westcountry by Downes, Jonathan
CFZ EXPEDITION REPORT: Gambia 2006 by Richard Freeman *et al*, Shuker, Karl (fwd)
The Owlman and Others by Jonathan Downes
The Blackdown Mystery by Downes, Jonathan
Big Cats in Britain Yearbook 2006 by Fraser, Mark (Ed)
Fragrant Harbours - Distant Rivers by Downes, John T
Only Fools and Goatsuckers by Downes, Jonathan
Monster of the Mere by Jonathan Downes
Dragons:More than a Myth by Freeman, Richard Alan
Granfer's Bible Stories by Downes, John Tweddell
Monster Hunter by Downes, Jonathan

Fortean Words

The Centre for Fortean Zoology has for several years led the field in Fortean publishing. CFZ Press is the only publishing company specialising in books on monsters and mystery animals. CFZ Press has published more books on this subject than any other company in history and has attracted such well known authors as Andy Roberts, Nick Redfern, Michael Newton, Dr Karl Shuker, Neil Arnold, Dr Darren Naish, Jon Downes, Ken Gerhard and Richard Freeman.

Now CFZ Press are launching a new imprint. Fortean Words is a new line of books dealing with Fortean subjects other than cryptozoology, which is - after all - the subject the CFZ are best known for. Fortean Words is being launched with a spectacular multi-volume series called *Haunted Skies* which covers British UFO sightings between 1940 and 2010. Former policeman John Hanson and his long-suffering partner Dawn Holloway have compiled a peerless library of sighting reports, many that have not been made public before.

Other books include a look at the Berwyn Mountains UFO case by renowned Fortean Andy Roberts and a series of forthcoming books by transatlantic researcher Nick Redfern. CFZ Press are dedicated to maintaining the fine quality of their works with Fortean Words. New authors tackling new subjects will always be encouraged, and we hope that our books will continue to be as ground-breaking and popular as ever.

Haunted Skies Volume One 1940-1959 by John Hanson and Dawn Holloway
Haunted Skies Volume Two 1960-1965 by John Hanson and Dawn Holloway
Space Girl Dead on Spaghetti Junction - an anthology by Nick Redfern
I Fort the Lore - an anthology by Paul Screeton
UFO Down - the Berwyn Mountains UFO Crash by Andy Roberts

"There's no sense in going further--it's the edge of cultivation,"
So they said, and I believed it--broke my land and sowed my crop--
Built my barns and strung my fences in the little border station
Tucked away below the foothills where the trails run out and stop:

Till a voice, as bad as Conscience, rang interminable changes
On one everlasting Whisper day and night repeated--so:
"Something hidden. Go and find it. Go and look behind the Ranges--
"Something lost behind the Ranges. Lost and waiting for you. Go!"

        Rudyard Kipling *The Explorer*

www.ingramcontent.com/pod-product-compliance
Lightning Source LLC
Chambersburg PA
CBHW070455100426
42743CB00010B/1626